Journal of Applied Logics - IfCoLog
Journal of Logics and their Applications

Volume 7, Number 1

January 2020

Disclaimer

Statements of fact and opinion in the articles in Journal of Applied Logics - IfCoLog Journal of Logics and their Applications (JALs-FLAP) are those of the respective authors and contributors and not of the JALs-FLAP. Neither College Publications nor the JALs-FLAP make any representation, express or implied, in respect of the accuracy of the material in this journal and cannot accept any legal responsibility or liability for any errors or omissions that may be made. The reader should make his/her own evaluation as to the appropriateness or otherwise of any experimental technique described.

College Publications
Scientific Director: Dov Gabbay
Managing Director: Jane Spurr

http://www.collegepublications.co.uk

EDITORIAL BOARD

SCOPE AND SUBMISSIONS

This journal considers submission in all areas of pure and applied logic, including:

pure logical systems
proof theory
constructive logic
categorical logic
modal and temporal logic
model theory
recursion theory
type theory
nominal theory
nonclassical logics
nonmonotonic logic
numerical and uncertainty reasoning
logic and AI
foundations of logic programming
belief change/revision
systems of knowledge and belief
logics and semantics of programming
specification and verification
agent theory
databases

dynamic logic
quantum logic
algebraic logic
logic and cognition
probabilistic logic
logic and networks
neuro-logical systems
complexity
argumentation theory
logic and computation
logic and language
logic engineering
knowledge-based systems
automated reasoning
knowledge representation
logic in hardware and VLSI
natural language
concurrent computation
planning

This journal will also consider papers on the application of logic in other subject areas: philosophy, cognitive science, physics etc. provided they have some formal content.

Submissions should be sent to Jane Spurr (jane@janespurr.net) as a pdf file, preferably compiled in LaTeX using the IFCoLog class file.

CONTENTS

Editorial
Special Issue on Multiple Valued Logic

Martin Lukac

Department of Computer Science, School of Engineering and Digital Sciences,
Nazarbayev University, Kazakhstan
`martin.lukac@nu.edu.kz`

While in general, Multiple-Valued Logic (MVL) saw its popularity relatively reduced, it still remains a valuable area for innovation, information exchange and growth. In addition the status of the MVL community and research has always been on the edge of current state of art technology nd applications. In fact many currently independent domains of research in computing are naturally multiple-valued and could be seen as being partially rooted in MVL. For instance, Fuzzy Logic is a special case of MVL. Quantum computing is also naturally multiple valued.

Indeed, quantum computing comes with a natural set of stable states useful for computation that is larger than two. While current quantum computing remains predominantly in the binary domain, multiple-valued quantum computing contains larger set of operators and states, resulting in larger possible saving in computation which is at this stage crucial for quantum supremacy.

This year the Multiple-Valued Logic (MVL) community continues producing exciting results in circuits, logic and design automation on the edge the current state of art. This special issue provides a glimpse of the current applications and research of the current state f art in MVL. Five articles on circuits and logic explore the advantages of using MVL in design and in theory of computation.

The first paper entitled *A Polarity-based Approach for Optimization of Multivalued Quantum Multiplexers with Arbitrary Single-qubit Target Gates* by Kevin Jin, Tahsin Saffat, Justin Morgan and Marek Perkowski, presents an approach for the design and optimization of multiple valued logic circuits. In this paper well known methods from classical logic design such as Reed-Muller representation and logic minimizations are used to design and optimized multiplexers intended to be used as multiple valued elementary quantum gates. Note that such circuits have been described in various approaches resulting in a general formulation of quantum circuit as a multiple-valued multiplexer.

A more traditional paper presents the contribution to the area of classical CMOS circuits, entitled *PAM-4 Signal Transmitter Using FPGA and DAC for Serial Link*

Test by Yasushi Yuminaka, Natsuki Sato, Takahito Chigira, Kohei toyoda and Yosuke Ijima. In this paper the four-level pulse amplitude modulation for ultra-high speed serial electrical interconnects is considered from the point of view of testing. The speeds of such transmission are reaching up to 400 Gbps. However the ultra high speed components requires a testing scheme: in the current state of art the testing are very expensive. The current paper proposes a multiple-valued transmitter using a DAC and FPGA for high speed and low cost serial ink testing supporting sophisticated digital equalization.

The next paper, is not directly related to multiple-valued logic but is on the edge of current computing. As its title indicates *Fast Hardware-Based Learning Algorithm for Binarized Perceptrons Using CMOS Invertible Logic* by Naoya Onizawa Duckgyu Shin and Takahiro Hanyu, the proposes to use CMOS invertible logic to build neural networks. Invertible logic is different from reversible logic in that it uses stochastic computing to allow forward and backward operations. This principle is generalized to the feed-forward neural networks that require both the forward pass for inference and backward pass for learning.

The next paper is a theoretical consideration and study of the para consistent quantum logic. The paper entitled *Some properties of first-order Nelsonian paraconsistent quantum logic* by Norihiro Kamide. The Nelsionian paraconsistent quantum logic is a quaternary paraconsistent quantum logic and in this paper important properties such as duality, conjunction or the negated existential quantifier are investigated and proved. Logic systems such as the one proposed are in par with the current logic systems used in modern computers and reasoning systems. Unlike in current systems multiple-valued logic is not a natural state of the computing devices and therefore it is not the first choice for computing. But in quantum computers the multiple-valued approach is natural and still a serious candidate for future main-stream quantum computing and logic.

The last paper in this special issue is dedicated to the study of content addressable memory and is entitled *Design of an MTJ-Based Nonvolatile Multi-Context Ternary Content-Addressable Memory* by Naoya Onizawa, Ren Arakawa and Takahiro Hanyu. This paper deals with a particular, ternary implementation of a content addressable memory. The particularity of this paper resides in the fact that the TCAD is designed using MTJ devices, a relatively new logical circuit elements allowing for non-volatile storage. Additionally, the combination of the MTJ with the CMOS logic circuits allows for high speed switching as well as a reduced area size.

These five papers are only a glimpse of the current MVL research and applications. While multiple-valued logic term is not as popular and main-stream as machine learning, multiple-valued logic is omnipresent in almost every filed and every application. From physics through math to bio-informatics, multiple-valued logic and reasoning is the exemplar tool to solve and to define problems.

Received October 2019

A Polarity-based Approach for Optimization of Multivalued Quantum Multiplexers with Arbitrary Single-qubit Target Gates

Kevin Jin
University of California, Berkeley
kevindujin@berkeley.edu

Tahsin Saffat
Massachusetts Institute of Technology
tahsin.saffat@gmail.com

Justin Morgan
Portland State University
jumorgan@pdx.edu

Marek Perkowski
Portland State University
h8mp@pdx.edu

Abstract

Previous work has provided methods for decomposing unitary matrices to series of quantum multiplexers, but the multiplexer circuits created in this way may be highly non-minimal. This paper presents a new approach for optimizing quantum multiplexers with arbitrary single-qubit quantum target functions and ternary controls. For multivalued quantum multiplexers, we define standard forms and two types of new forms: Fixed Polarity Quantum Forms (FPQFs) and Kronecker Quantum Forms (KQFs). Drawing inspiration from the usage of butterfly diagrams, we devise a method to exhaustively construct new forms. In contrast to previous butterfly-based methods, which are used with classical Boolean functions, these new forms are used to optimize quantum circuits with arbitrary target unitary matrices. Experimental results on the new forms applied to various target gates such as NOT, V, V$^+$, Hadamard, and Pauli rotations, demonstrate that these new forms greatly reduce the gate costs of ternary quantum multiplexers.

1 Introduction

Previous works in the field of quantum compilation, such as [1, 22], have generated methods for decomposition of arbitrary unitary matrices into a series of quantum multiplexers, but the multiplexers created by these methods may be highly non-minimal. Several papers are related to the synthesis of binary quantum circuits with controlled V and V^+ gates [17, 18, 20, 21, 22, 23, 24, 25, 26, 27]. Although quantum computers now allow for multivalued logic, very little has been published on circuit synthesis with ternary or higher-order multivalued multiplexers. Our method addresses the case where the gates of the multiplexers are controlled by many ternary inputs, but target only a single binary qubit. For this case, our results are significantly less expensive, or even exactly minimal. This paper, by inventing the new concept of polarity-based multiplexers with arbitrary targets, gives a new methodology for optimization of ternary-input, binary-output quantum circuits. Because there are no benchmarks for these kinds of quantum circuits, we use randomly generated benchmark data and some examples from previous papers. We also briefly introduce the concept of ternary-controlled multiplexers with ternary target gates (ternary-input, ternary-output), but our numerical results are restricted to ternary-input, binary-output multiplexers.

Section 2 gives necessary background for our paper, introducing previous research and concepts. Section 3 introduces new multi-valued multiplexer forms, Fixed Polarity Quantum Forms (FPQFs), and Kronecker Quantum Forms (KQFs), which generalize the ideas from binary Reed Muller forms to forms that control arbitrary single-qubit gates. This section also includes an explanation of why our butterfly decompositions work for transforming standard form multiplexers to FPQFs and KQFs. Section 4 discusses the program used to compute matrix transformations and FPQF/KQF forms and also introduces how costs were calculated for these new types of generalized multiplexers. Section 5 analyses and discusses the results generated by our program. Because the best improvements for general multiplexers were found for multiplexers that control V, V^+, and NOT gates, we analyze these results as a special case. Such circuits are created as intermediate results in the methods from previous research discussed in Section 2. Section 6 draws conclusions and summarizes the paper.

2 Background

2.1 General background on quantum multiplexers and their place in quantum circuit synthesis

The **quantum multiplexer** is an important concept in quantum circuit synthesis. In previous literature, it was related to multi-qubit control of target gates, which are theoretically arbitrary. However, most of the earlier work on quantum multiplexers was related to multi-controlled-NOT gates, such as Toffoli. For instance, this was done for the synthesis of ESOP circuits as well as some special cases of ESOP, such as Fixed Polarity Reed-Muller (FPRM) and Kronecker Reed-Muller (KRM). Very little has been published on optimizing similar ternary-controlled forms and circuits; a rare example is [28]. While some methods decompose arbitrary unitary matrices to smaller binary-controlled quantum multiplex-

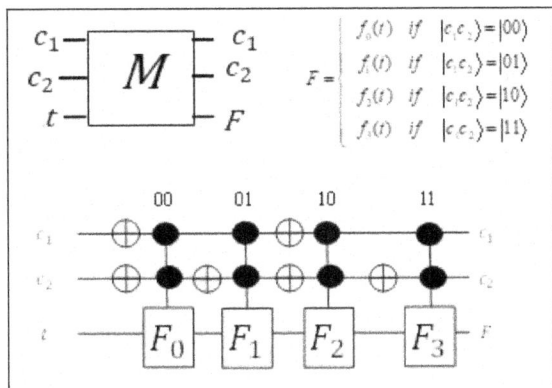

Figure 1: An arbitrary binary standard form quantum multiplexer with two controls, represented both mathematically and as a circuit diagram. The controls are listed above each function.

ers [1, 2], only [22] covers the ternary-controlled case. In our paper, we are not interested in this initial decomposition stage, nor are we interested in the specific optimization of ternary counterparts of classical Reed-Muller-like forms. Rather, we introduce a new concept of butterfly-like diagrams to optimize ternary-controlled quantum multiplexers with arbitrary target single-qubit gates; this is in contrast to multiplexers in previous work [14, 15], where multi-controlled, binary-controlled gates (such as Toffoli) only control NOT gates as target.

The concept of a more general quantum multiplexer was first introduced by Shende et. al. in [3], where they propose the quantum multiplexer circuit block for usage in recursive decompositions of arbitrary unitary matrices. Other work by Shende et. al. in [4] focused on optimization of two-qubit unitary operators, but not on optimization of larger circuits such as quantum multiplexers. In [2], Vartiainen et. al. demonstrate a method for optimization of arbitrary multi-qubit gates, but do not provide a method specifically for optimization of the less general multi-qubit multiplexers. Tucci developed the Qubiter program in [1] to decompose arbitrary unitary matrices to series of quantum multiplexers, but a later work by Hutsell [9] has shown that the output from Qubiter is highly non-minimal, leaving plenty of space for optimization. There are high degrees of similarities between lookup-tables (LUTs) and multiplexers, but the multiplexers introduced by us have arbitrary target gates, in contrast with LUTs, which only have NOTs as target gates. Work by Soeken et. al. [14, 15] mapped networks of LUTs to reversible gates; our work creates large LUTs, controlling arbitrary targets, to realize the entire function. Work by Gidney [13] reduced the cost of adder circuits by reducing the number of T Gates required, while our method allows for arbitrary target gates, although we did not use T Gates.

Yet another application of quantum multiplexers is not to control inverters only, but rather gates that are square roots of NOTs, such as V and V$^+$, and also other higher-order roots of NOT and their Hermitians [17, 18, 20, 21, 22, 23, 24, 25, 26, 27]. The papers in this area create circuits with multi-controlled gates that can be optimized. For example, a Controlled-V gate composed with a Controlled-V$^+$ gate creates a two-qubit Identity, which means that these two gates can be removed from the circuit.

In this paper, we propose a method that will take into account all the aforementioned applications of quantum multiplexers.

2.2 Multivalued logic

A q-valued quantum dit is a qudit with q different basis states, which we will call $|0\rangle, |1\rangle, \ldots, |q-1\rangle$. A set of m q-valued qudits can occupy q^m different basis states, and their collective state is a superposition of these basis states.

The single qudit gates for q-valued logic can be represented by arbitrary permutative $q \times q$ matrices. For instance, in ternary logic, these quantum gates are $(+1), (+2), (01), (02)$, and (12), where:

$$(+1) = \begin{bmatrix} 0 & 0 & 1 \\ 1 & 0 & 0 \\ 0 & 1 & 0 \end{bmatrix}; \qquad (+2) = \begin{bmatrix} 0 & 1 & 0 \\ 0 & 0 & 1 \\ 1 & 0 & 0 \end{bmatrix}$$

The other ternary gates are not important to our paper. Cyclic inverters are an important type of multivalued gates. These are generalizations of the NOT for multi valued quantum logic. We define the cyclic inverter r_k as the gate that takes a qubit in the state $|a\rangle$ to the state $|a + k\rangle$, where the addition is taken modulo q. In the case of ternary, $(+1)$ and $(+2)$ are both cyclic inverters.

8

2.3   Binary quantum multiplexers and controlled gates

For didactic reasons, let us first define the binary quantum multiplexer. A **binary quantum multiplexer** is a block of gates that will, based on a set of binary-valued control variables (qubits), use a Boolean product of the control literals to select an arbitrary unitary binary-valued quantum function (called a **target function**, or a **target gate**) that acts on a single target variable (qubit): see Figure 1. Note that in the diagram, the black circles denote âĂIJcontrolsâĂİ. That is, if the signal on the line is $|1\rangle$ when it reaches the control, then the control is enabled. If all controls of a particular target gate are enabled, then the gate will activate. Thus, when taken together, the controls of a target function form its **control function** (the previously mentioned product of control literals). For example, the function F_0 is controlled by both $\bar{c}_1$ and $\bar{c}_2$, which means that the control function is $\bar{c}_1\bar{c}_2$, and the target function is active if and only if $c_1 = 0$ and $c_2 = 0$. Target functions can be any single-qubit quantum functions. Thus, they are represented by unitary 2×2 matrices.

Suppose we have a multiplexer M with m control variables, which we can denote as $c_1, c_2, \ldots, c_m$. We can define any of the possible input states to the multiplexer in terms of the values on the control variables: the **input state** i can be represented as a binary string, where each digit i_k is equal to the value of its corresponding control variable c_k. For example, if we have c_1, c_2, c_3 which have values of $0, 1, 1$ respectively, i would be $011 = 3$. Note that there are only 2^m possible input states since we are assuming that the control variables are always in basis quantum states. However, we can be dealing with superposition states in the target qubit.

For the multiplexer M, let F_i be the arbitrary single-qubit quantum function that will act on the target variable if and only if the input state is i: we can say that the function F_i is a **target (controlled) function** that is **controlled by** i. For example, for the function $F_3 = F_{011}$ controlled by the three variables c_1, c_2, c_3, we say that F_3 is controlled by $i = 3 = 011$; this is equivalent to saying that F_3 is controlled by the Boolean product $\bar{c}_1 c_2 c_3$ since only the input state $i = 3$ satisfies this control function. It is clear that any (binary) multiplexer with m control variables can be represented as 2^m controlled functions, one for each input state. Thus, we can uniquely represent the multiplexer as an ordered set:

$$M = \{F_0, F_1, \ldots, F_{2^m-1}\}_C$$

where C is the ordered set of control variables. We choose to represent the multiplexer in this fashion for mathematical convenience in the âĂIJProofâĂİ section below.

Note that a multiplexer of this form is a direct realization of a Minterm Sum of

Products form: each minterm of the control variables controls its own target function. For example, for a multiplexer with two control variables c_1, c_2, the minterms of the controls $\bar{c}_1\bar{c}_2, \bar{c}_1c_2, c_1\bar{c}_2, c_1c_2$ each appear once and each control a separate function F_0, F_1, F_2, F_3, respectively (once again, see Figure 1). Henceforth, this form of multiplexer that mirrors the Minterm Sum of Products form will be referred to as the **standard form**.

2.4 Multivalued quantum multiplexers

A multivalued quantum multiplexer is a gate that will, based on a set of q-valued control variables, use a classical q-valued function to select an arbitrary binary target function that acts on a set of target variables. These target functions can be any quantum functions, acting on any number of target variables. Thus, they are represented by unitary matrices, and the entire quantum multiplexer can be represented by a âĂIJunitary block matrixâĂİ. Figure 2 gives a general ternary multiplexer.

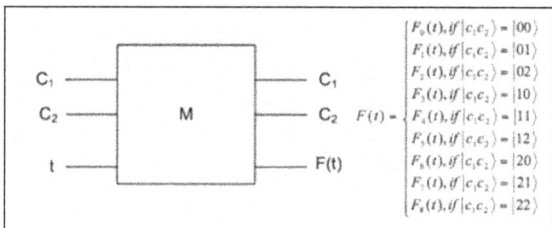

Figure 2: An arbitrary ternary standard form quantum multiplexer with two ternary-valued controls.

Suppose we have a multiplexer M with m q-valued control variables, which we will denote $c_1, c_2, \ldots, c_m$. We can represent an input state as a base q string of numbers, where each digit i_k is equal to the value of its corresponding control variable c_k. Note that there are only q^m possible inputs (input states) since we are assuming that the control variables are always in basis states. However, we will be dealing with superposition states of the target variables.

For the multiplexer M, let F_i be the unitary matrix that will act on the target variables if the input is i. Also, let C be the set of control variables. Then we can uniquely represent the multiplexer as an ordered set:

$$M = \{F_0, F_1, \ldots, F_{q^m-1}\}_C$$

In this paper, the multivalued multiplexers that we predominately deal with will have ternary-valued control variables and binary-valued target functions.

10

3 Fixed polarity quantum forms (FPQF) and Kronecker quantum forms (KQF)

3.1 Introduction to multivalued FPQF and KQF

One may be familiar with the Fixed Polarity Reed Muller form (FPRM), a polarized, canonical form of a binary Boolean expression where each variable appears solely in either its complemented (negative polarity) or uncomplemented (positive polarity) form. For a function on m variables, there are 2^m different FPRM forms since each of the variables can either be complemented or uncomplemented

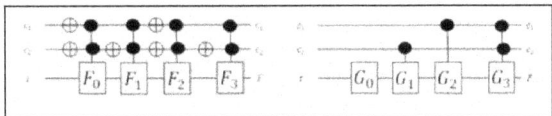

Figure 3: Circuit diagrams for binary multiplexers. The left diagram shows a standard form multiplexer; the right diagram shows a FPQF multiplexer in polarity 11.

(there are 2 choices for each of the m variables). FPRM is related to Positive Davio and Negative Davio expansions. There is also the Kronecker Reed Muller form (KRM), a canonical form of a binary Boolean expression where each variable appears in either positive, negative, or mixed polarity. There are 3 choices (positive, negative, or mixed) for each variable, so there are 3^m different KRMs for a function on m variables. FPRM is a subset of KRM since KRM is related to Positive Davio and Negative Davio expansions, along with the Shannon expansion. If one is familiar with these Boolean forms, it may be easier to understand our new forms, but knowledge of FPRM and KRM is not critical to understanding the new concepts that we introduce.

For quantum multiplexers, given a set of control variables, $c_1, c_2, \ldots, c_m$, we can introduce the concept of polarity in a similar fashion as it appears in FPRM or KRM. We can do so by creating a multiplexer where each control variable is in a fixed polarity (either positive or negative polarity); such a multiplexer will henceforth be referred to as a **Fixed Polarity Quantum Form (FPQF)**. Similarly, we can create a multiplexer where control variables are in either fixed or mixed polarity; such a multiplexer will be referred to as a **Kronecker Quantum Form (KQF)**. When a multivalued variable is in fixed polarity, the control that is activated by a check against $|0\rangle$ becomes implicit and is no longer present; for example, for a ternary control variable in fixed polarity, controls will only be placed to check for the values of $|1\rangle$ and $|2\rangle$, but not $|0\rangle$ ($|0\rangle$ will be implicit, and any target functions that originally relied on a check against $|0\rangle$ will instead disregard that control variable completely). Figure 3 contrasts the binary-valued standard form with an FPQF form. If we wish to create an FPQF or KQF multiplexer that realizes exactly the

11

same functions as a particular standard form multiplexer for all input states while fixing the polarities of some (or all) of the control variables, it is clear that the functions $[G_0, G_1, \ldots, G_{q^m-1}]$ of this FPQF or KQF multiplexer will be different from the functions $[F_0, F_1, \ldots, F_{q^m-1}]$ of the standard form multiplexer since the control functions have changed: see Figure 5 for an example of the way that the control functions change. In an FPQF multiplexer, it is possible for multiple target functions to be active for a given input state; for example, for the input state 10 to a binary multiplexer, both G_0 and G_2 are active.

Henceforth, we will denote target functions as F_i if they are the target functions in a standard form multiplexer, and we will denote them as G_i if they are the target functions of an FPQF/KQF multiplexer. First, it is useful to formally define how the target functions of a FPQF multiplexer G_i operate.

In an FPQF multiplexer, G_i is a unitary matrix acting on the target variable and controlled by all input states j such that $j_k = i_k$ if $i_k \neq 0$ (where j_k and i_k are the kth digits of j and i, respectively). That is, G_i is only controlled by the control variable c_k if the kth digit of i is not 0. We have previously mentioned that multiple target functions may be active for a particular input state; conversely, a target function can be activated by multiple input states. For example, on a binary-valued multiplexer with two controls, $G_1 = G_{01}$ is controlled by both the input states 01 and 11. That is, G_1 is activated if the input state is 01 *or* 11. Similar to our notation for a standard

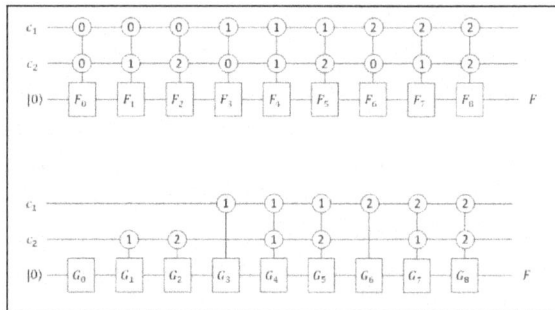

Figure 4: Circuit diagrams for ternary multiplexers. The upper diagram shows a standard form multiplexer; the lower diagram shows a FPQF multiplexer in polarity 22. (polarity 22 means that both variables are in polarity 2)

form multiplexer, we denote an FPQF multiplexer as $M = \{G_0, G_1, \ldots, G_{q^m-1}\}_{C,p}$ where p is the polarity of the FPQF multiplexer as represented by a base q string. For q-valued logic, there are q^m possible FPQF forms for a given set of functions F_i, so we need a fast way to compute all possible polarities of $[G_0, G_1, \ldots, G_{q^m-1}]$ from $[F_0, F_1, \ldots, F_{q^m-1}]$. Note that the circuit structures for all FPQF polarities are highly similar; while Figure 4 shows an FPQF multiplexer in polarity 22, other polarities can be realized simply by applying a cyclic inverter to the beginnings of control lines that we would like to place in an alternative polarity, then modifying the target functions G_i.

12

In a KQF multiplexer, whether or not a gate G_i is active is dependent on the polarity. If the kth variable is in a fixed polarity, G_i is controlled by all input states j such that $j_k = i_k$ if $i_k \neq 0$. That is, unless i_k is 0, G_i is controlled by the control variable c_k. If the kth variable is in a mixed polarity, G_i is controlled by all input states j such that $j_k = i_k$. Alternate fixed polarities can be realized simply by applying a cyclic inverter to the beginnings of controls lines that we would like to place in a different polarity, then modifying the target functions G_i accordingly. Figure 4 gives two example circuit diagrams for ternary multiplexers. The notation for controls is slightly different for ternary multiplexers. The control circles contain numbers. If the signal of the line is equal to the contained number when it reaches the control, then the control is enabled; otherwise, it is not.

There is a clear mathematical relationship between the target functions F_i of the standard form multiplexer and the target functions G_i of the FPQF multiplexer. Below we provide an example of how this relationship can be calculated.

$$
\begin{array}{ll}
F_0 = G_0 & G_0 = F_0 \\
F_1 = G_1 \cdot G_0 & G_1 = F_1 \cdot F_0^{-1} \\
F_2 = G_2 \cdot G_0 & G_2 = F_2 \cdot F_0^{-1} \\
F_3 = G_3 \cdot G_2 \cdot G_1 \cdot G_0 \quad\Longrightarrow\quad & G_3 = F_3 \cdot F_1^{-1} \cdot F_0 \cdot F_2^{-1}
\end{array}
$$

Figure 5: Mathematical relationships between the target functions of a binary standard form multiplexer F_i and the target functions of an FPQF multiplexer G_i in polarity 11. This example case involves multiplexers with two control variables (and thus four target functions).

Remember that a gate F_i or G_i is active if all its controls are enabled. Consider an example with two binary control variables, where we want the standard form and FPQF multiplexers to realize exactly the same target functions for all different control inputs. For input state 00, F_0 is active in the standard form multiplexer and G_0 is active in the FPQF multiplexer. Thus, $F_0 = G_0$. For input state 01, F_1 is active in the standard form multiplexer, but two gates are active in the FPQF multiplexer: G_0, G_1 (Figure 3 may help to visualize this). Thus, $F_1 = G_1 \cdot G_0$, and therefore $G_1 = F_1 \cdot G_0^{-1} = F_1 \cdot F_0^{-1}$. Note that the $\cdot$ operation is the matrix multiplication operation applied on the target functions when represented as unitary matrices. Similar observations can be made for input states 10 and 11; see Figure 5 for all the mathematical formulas that we derive. We have shown that the functions G_i can be directly calculated from the functions F_i for an example with two controls. Next, we demonstrate how to compose a transformation that can perform these calculations, which will allow us to generalize the process of calculating G_i from F_i.

For a given multiplexer $M = \{F_0, F_1, \ldots, F_{q^m-1}\}_C$ and a given KQF form $M = \{G_0, G_1, \ldots, G_{q^m-1}\}_{C,p}$, define $T_{m,p}$ as the transformation over U, the space of valid $m \times m$ unitary matrices, that takes the vector $[F_0, F_1, \ldots, F_{q^m-1}]$ to the vector

$[G_0, G_1, \ldots, G_{q^m-1}]$ for a certain polarity p (remember that polarities of a multiplexer are represented as base q strings, where the ith digit denotes the polarity for the ith control variable, in the same way as we previously introduced for FPRM polarities in [29]). We now provide a formal explanation for decomposing $T_{m,p}$ into a series of simpler transformations, which we can then represent with KQF butterfly diagrams similar to how it is used in classical KRM butterflies [5, 10].

3.2   The Kronecker product

Figure 6 illustrates the concept of the Kronecker product. The Kronecker product can be applied to any two matrices, each of arbitrary dimension.

$$\begin{bmatrix} a & b \\ c & d \end{bmatrix} \otimes \begin{bmatrix} x & y \\ z & w \end{bmatrix} = \begin{bmatrix} ax & ay & bx & by \\ az & aw & bz & bw \\ cx & cy & dx & dy \\ cz & cw & dz & dw \end{bmatrix}$$

Figure 6: Example of Kronecker product ($\otimes$) on two 2×2 matrices

3.3   Formal explanation of multivalued butterfly decomposition

This explanation demonstrates that we are able to decompose the transformation that takes a standard form multiplexer to a KQF multiplexer into a series of butterfly diagrams.

We first discuss a notation used in the following proof. Let us define an arbitrary transformation B, which acts on a subset of all the functions of a multiplexer. We define the butterfly transformation $B_{n,i}$ as a transformation that acts on a set of q^n functions represented as unitary matrices, $[F_0, F_1, \ldots, F_{q^n-1}]$, such that every set of functions $F_{r0}, F_{r1}, \ldots$, where the base q representations of $r0, r1, \ldots$ differ only at the ith digit, is acted upon by the transformation B. For example, if B takes $[a, b]$ to $[b, a \cdot b]$ for a **binary** case, then $B_{2,1}$ is the transformation performed on $2^2 = 4$ functions, where B is acted upon pairs of functions whose binary representations of index only differ at the 1st digit. The transformation $B_{2,1}$ would take the functions $[F_0, F_1, F_2, F_3]$ to $[F_1, F_0 \cdot F_1, F_3, F_2 \cdot F_3]$. This is because B was applied to the pairs $[F_0, F_1] = [F_{00}, F_{01}]$ and $[F_2, F_3] = [F_{10}, F_{11}]$ since the indices of each pair of functions differ only at the 1st digit. As an additional example, if B takes $[a, b, c]$ to $[b, c \cdot b^{-1}, a \cdot b^{-1}]$ for a ternary case, then $B_{2,1}$ is the transformation performed on $3^2 = 9$ functions, where B is acted upon the triplets of functions whose ternary representations of index only differ at the 1st digit. The transformation $B_{2,1}$ would take the functions $[F_0, F_1, F_2, F_3, F_4, F_5, F_6, F_7, F_8]$ to $[F_1, F_2 \cdot F_1^{-1}, F_0 \cdot F_1^{-1}, F_4, F_5 \cdot F_4^{-1}, F_3 \cdot F_4^{-1}, F_7, F_8 \cdot F_7^{-1}, F_6 \cdot F_7^{-1}]$ since B is

14

applied to the triplets $[F_0, F_1, F_2]$, $[F_3, F_4, F_5]$, and $[F_6, F_7, F_8]$. See Figure 9 for an illustration of this particular ternary transformation. Note that $B_{n,i}$ is the algebraic representation of a column of butterflies on q^n objects where each butterfly kernel is stretched by q^i; thus, we call $B_{n,i}$ a butterfly transformation. A more in-depth explanation of FPRM and KRM butterflies (which are useful, but not essential, to understand our FPQF and KQF butterflies) can be found in [29]. We now show how a KQF transformation for q-valued controls can be decomposed to butterfly transformations: $A_0, A_1, \ldots, A_{q-1}$.

First, we define the transformations A_p, for $0 \leq p \leq q - 1$. We define A_p as a degree q transformation that takes $[a_0, a_1, \ldots, a_{q-1}]$ to $[b_0, b_1, \ldots, b_{q-1}]$ such that $b_0 = a_{p+1}$ and $b_k = a_{k+p+1} \cdot a_{p+1}^{-1}$ for $k \neq 0$. For example, for ternary (degree $q = 3$), A_0 would take $[a_0, a_1, a_2]$ to $[b_0, b_1, b_2] = [a_1, a_2 \cdot a_1^{-1}, a_0 \cdot a_1^{-1}]$, A_1 would take $[a_0, a_1, a_2]$ to $[b_0, b_1, b_2] = [a_2, a_0 \cdot a_2^{-1}, a_1 \cdot a_2^{-1}]$, and A_2 would take $[a_0, a_1, a_2]$ to $[b_0, b_1, b_2] = [a_0, a_1 \cdot a_0^{-1}, a_2 \cdot a_0^{-1}]$. We claim that $T_{m,p} = A_{p_m} \otimes A_{p_{m-1}} \otimes \ldots \otimes A_{p_1}$. We prove this by induction on m.

3.3.1 Base Case: We first establish the claim for $m = 1$

We start by calculating $T_{1,p}$. We have $M = \{F_0, F_1, \ldots, F_{q-1}\}_C = \{G_0, G_1, \ldots, G_{q-1}\}_{C,p}$. In order for this to be true, the multiplexers must output the same values for all inputs of the control variable. By definition, for the multiplexer $\{G_0, G_1, \ldots, G_{q-1}\}_{C,p}$, the function G_k for $k \neq 0$ is selected by k and the polarity shifts the control variable by $q - 1 - p$, so G_k is selected when the control variable is $(k - (q - 1 - p)) \bmod q = (k + p + 1 - q) \bmod q = k + p + 1$. Thus, we have $G_0 = F_{p+1}$. For $k \neq 0$, we have $G_k \cdot G_0 = F_{k+p+1}$, so $G_k = F_{k+p+1} \cdot G_0^{-1} = F_{k+p+1} \cdot F_{p+1}^{-1}$. Thus, the transformation A_p over U takes $[F_0, F_1, \ldots, F_{q-1}]$ to $[G_0, G_1, \ldots, G_{q-1}]$. Thus, we have $T_{1,p} = A_p$ as desired.

3.3.2 Induction: We establish the claim for m, assuming it holds for $m - 1$.

For the induction step it suffices to show that $T_{m,p} = T_{m-1,p'} \otimes T_{1,p_1}$, where $p' = p - p_1 q^{m-1}$ (p' is the polarity p without its leading digit, p_1).

Given a multiplexer $M = \{F_0, F_1, \ldots, F_{q^m-1}\}_C$ with control variables $c_1, c_2, \ldots, c_m$ and t target variables, it can be reinterpreted as a multiplexer with one control variable, c_1, and $t + m - 1$ target variables. If we let C' be the set of control variables without c_1, then $M = \{M_0, M_1, \ldots, M_{q-1}\}_{c_1}$, where each of M_k is a multiplexer of $m - 1$ control variables that can be expressed as $M_k = \{F_{kq^{m-1}}, F_{kq^{m-1}+1}, \ldots, F_{(k+1)q^{m-1}-1}\}'_C$. See Figure 7 for an illustration of this.

Now, in order to show that $T_{m,p} = T_{m-1,p'} \otimes T_{1,p_1}$, we will realize M in two steps. First, we implement M as a multiplexer with one control variable in KQF form. That is:

$$M = \{M_0', M_1', \ldots, M_{q-1}'\}_{c_1, p_1}$$

We will then show that each of M_k' is a multiplexer, so we will decompose each of the M_k' into KQF form with polarity p'.

First, we show that each of M_k' is a multiplexer. Note that the composition of two multiplexers is another multiplexer with target functions that are the composition of the original multiplexersâĂŹ target functions. And the inverse of a multiplexer is another multiplexer whose target functions are the inverses of the original multiplexerâĂŹs target functions. Thus, since the transforma-

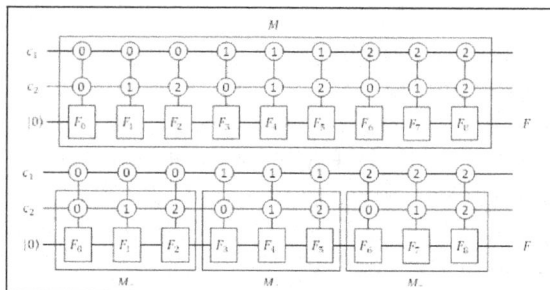

Figure 7: Example illustration of the reinterpretation of a multiplexer as a series of smaller multiplexers.

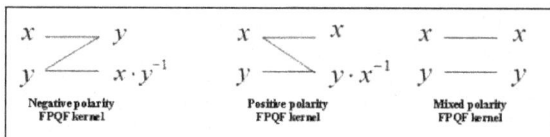

Figure 8: Butterfly kernels for binary FPQF and KQF butterfly diagrams. Only the first two types of kernels are present in FPQF diagrams.

tion T_{1,p_1} takes $[M_0, M_1, \ldots, M_{q-1}]$ to $[M_0', M_1', \ldots, M_{q-1}']$ each of the functions M_k' is a multiplexer. Now, let $M_k' = \{F_{kq^{m-1}}', F_{kq^{m-1}+1}', \ldots, F_{(k+1)q^{m-1}-1}'\}_{C'}$ Then, we can easily compute $[F_0', F_1', \ldots, F_{q^m-1}']$ from $[F_0, F_1, \ldots, F_{q^m-1}]$. Once again, because of the way multiplexers compose, the transformation T_{1,p_1} also takes $[F_k, F_{k+q^{m-1}}, F_{k+2q^{m-1}}, \ldots, F_{k+(q-1)q^{m-1}}]$ to $[F_k', F_{k+q^{m-1}}', F_{k+2q^{m-1}}', \ldots, F_{k+(q-1)q^{m-1}}']$ for all $0 \leq k \leq q^{m-1} - 1$.

Finally, we can realize $M = \{G_0, G_1, \ldots, G_{q^m-1}\}_{C,p}$ from $M = \{M_0', M_1', \ldots, M_{q-1}'\}_{c_1, p_1}$ by representing each of the multiplexers M_k' in KQF form. Thus, we have:

$$M_k' = \{F_{kq^{m-1}}', F_{kq^{m-1}+1}', \ldots, F_{(k+1)q^{m-1}-1}'\}_{C'}$$

$$= \{G_{kq^{m-1}}, G_{kq^{m-1}+1}, \ldots, G_{(k+1)q^{m-1}-1}\}_{C', p'}$$

16

Thus, the transformation $T_{m-1,p'}$ takes $[F'_{kq^{m-1}}, F'_{kq^{m-1}+1}, \ldots, F'_{(k+1)q^{m-1}-1}]$ to $[G_{kq^{m-1}}, G_{kq^{m-1}+1}, \ldots, G_{(k+1)q^{m-1}-1}]$. Thus, by the definition of the Kronecker product, $T_{m-1,p'} \otimes T_{1,p_1}$ takes $[F_0, F_1, \ldots, F_{q^m-1}]$ to $[G_0, G_1, \ldots, G_{q^m-1}]$ and we have $T_{m,p} = T_{m-1,p'} \otimes T_{1,p_1}$ as desired.

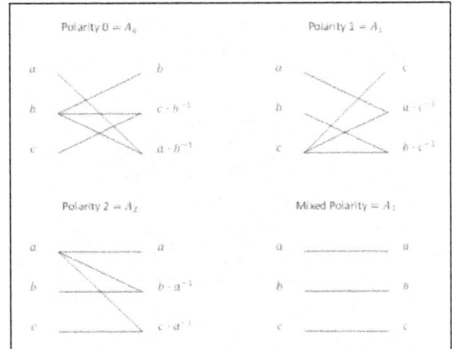

Figure 9: Butterfly kernels for ternary FPQF and KQF butterfly diagrams. Only the first three types of kernels are present in FPQF diagrams.

4 Discussion of the program used to compute FPQF and KQFs

4.1 FPQF and KQF butterflies

Here, we introduce the FPQF and KQF butterfly kernels that are used in FPQF and KQF butterflies. These are illustrated in Figure 8 for binary, and Figure 9 for ternary.

They are based on the previously given explanation for composition. In Figure 9, the four butterflies correspond to A_0, A_1, and A_2, along with a special case A_3 for mixed polarity in KQF, which is a transformation that has no effect on the target functions. The appearance of the binary kernels in Figure 8 may seem arbitrary, but they are in fact based on the appearance of butterfly kernels for FPRM butterfly diagrams and other Fast Fourier Transform butterflies .

The construction of FPQF butterfly diagrams is similar to the construction of FPRM butterfly diagrams, as discussed in [29]. However, note that the inputs to the diagram are no longer the minterms of a Boolean function, and the outputs are no longer spectral coefficients; rather, the inputs are the target functions F_i of the standard form multiplexer, and the outputs are the target functions G_i of the polarized multiplexer. See Figure 10 for an example of the FPQF butterfly diagrams for all polarities of FPQF on a binary multiplexer with three controls . Figure 11 gives an example FPQF butterfly diagram for a single polarity of a ternary multiplexer with two controls , and Figure 12 gives further examples. Figure 12 also demonstrates that the target functions have no effect on the shape of the butterfly kernels; only the polarity has an effect.

17

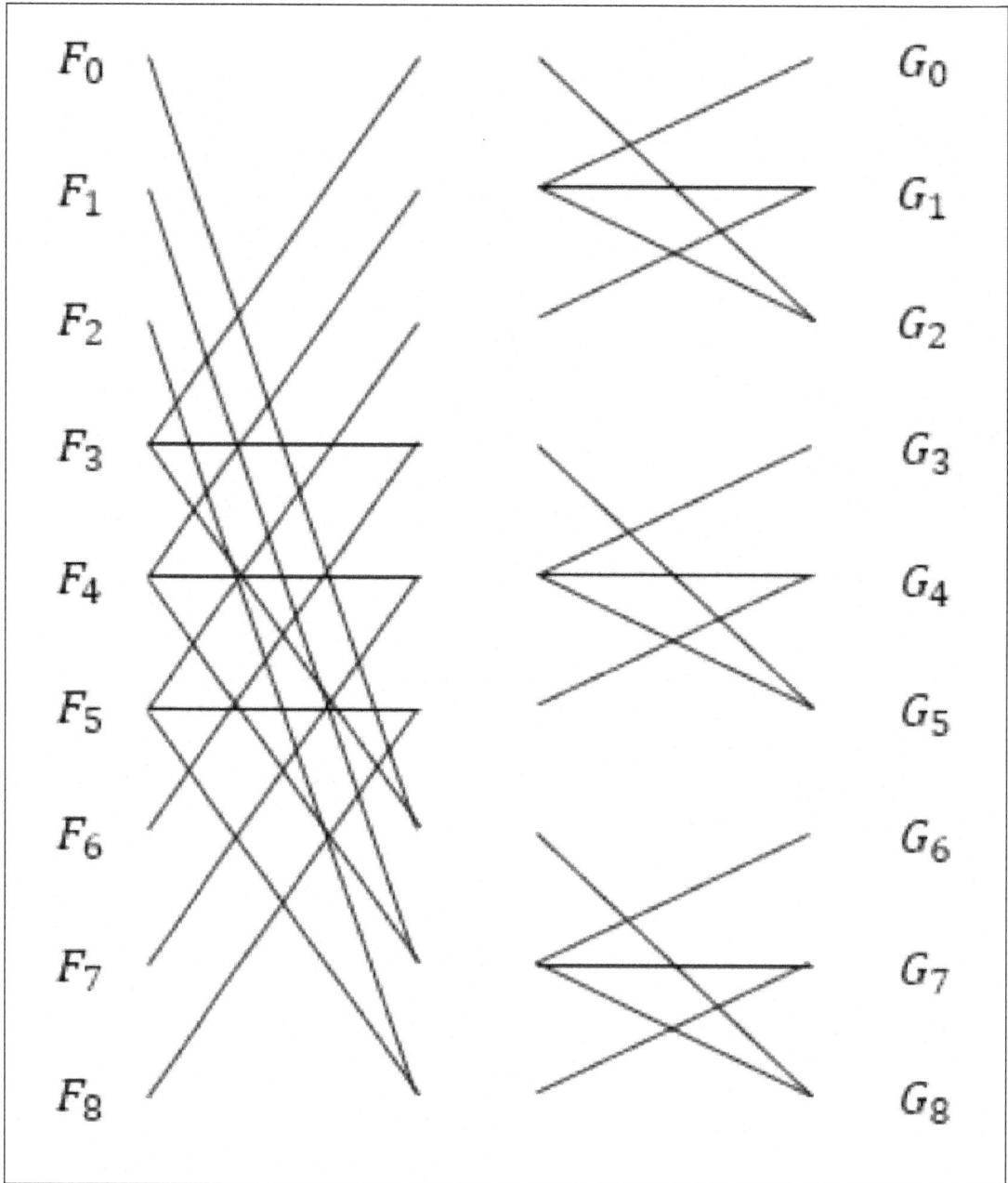

Figure 11: Sample KQF butterfly diagram for a standard form multiplexer with two ternary control variables. This butterfly converts a ternary standard form multiplexer into a KQF form with polarity 00.

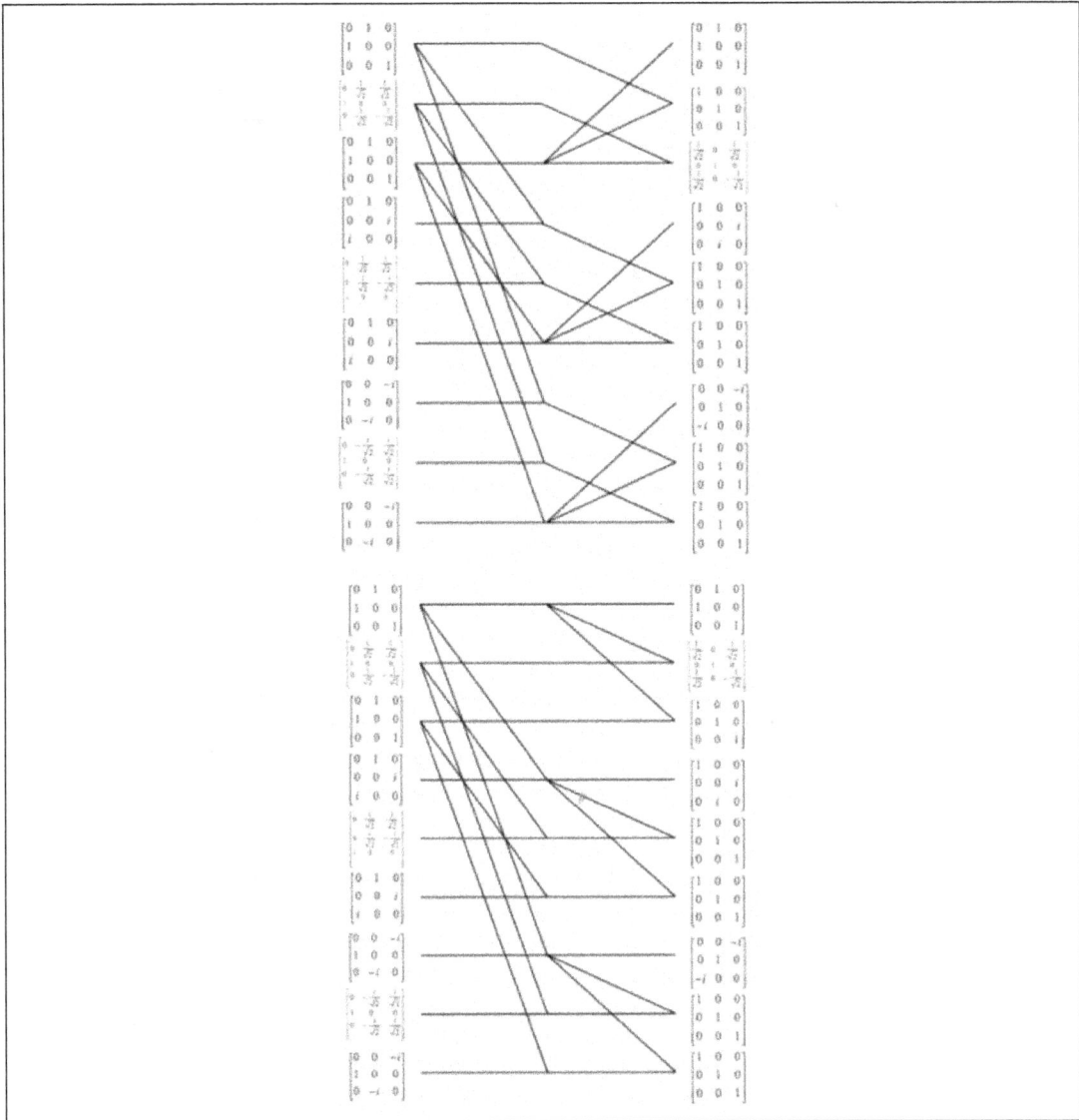

Figure 12: Two arbitrary examples of butterfly diagrams. These diagrams are for multiplexers with ternary-input, ternary-output; in contrast, our program functions for ternary-input, binary-output.

4.2   Quantum cost of FPQF/KQF polarities

Having developed a method to calculate all possible FPQF and KQF forms
of a multivalued standard form multiplexer (that is, by constructing and
evaluating all possible FPQF or KQF
butterfly diagrams), we now introduce
how the cost of quantum multiplexers will be calculated for our ternary
multiplexer program. Similar to [8],
we define the cost of a quantum circuit in terms of the number of uncon-
trolled and single-controlled gates required to realize it; we can find this
by summing the costs of all the controlled gates in the multiplexer circuit.
The concept is being discussed with
regards to ternary-valued multiplexers;
however, formulations for the costs of
multi-controlled ternary gates do not
exist yet. Thus, we approximate cost
by using currently existing costs for
multi-controlled binary gates. Maslov
et.al. [7] has previously established the

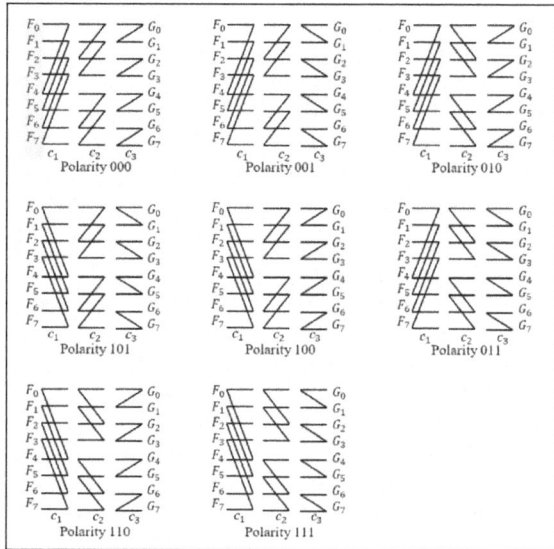

Figure 10: All polarities of FPQF butterflies for a standard form binary multiplexer with three control variables. Note
that the sequences of functions G_i generated by each FPQF butterfly are different.

following cost functions for binary Toffoli gates with m controls (see Table 1). We
will assume that we are able to use a single Ancilla bit for our entire circuit, so
the equation $32m - 96$ is relevant to us. For reasons that are not important to the
understanding of this paper (and which are discussed in [8]), it is therefore theoretically possible to create any multi-controlled gate through the same process as the
construction of a Toffoli: see *Lemma 6.1* in [8]. Thus, the cost functions that have
been developed by [8] for Toffoli gates can also be applied to formulate the cost of
multi-controlled binary gates of any type.

For approximation, we will assume that the cost of multi-controlled ternary gates
with n controls is directly proportional to the cost of multi-controlled binary gates
with n controls. Additionally, we assume that the cost of cyclic inverters is similar to
the cost of a binary inverter; in other words, we assume that$(+1)$ and $(+2)$ gates cost
roughly as much as a NOT gate. We base this assumption on results from previous
work by Wang et. al. [28], where multi-controlled Toffoli-like gates were realized

20

Multi-controlled gate size $(m+1)$	Number of Ancilla bits	Gate Cost
1	0	1
2	0	1
3	0	5
4	0	13
5	0	29
6	1	52
7	1	84
8	1	116
9	1	154
10	1	192
$(m+1) > 10$	1	$32m - 96$

Table 1: Costs of multi-controlled gates with m controls. These gates use one or fewer Ancilla bits, which can be reused. Reproduced from [8]

for ternary logic with costs that were proportional to multi-controlled binary Toffoli gates.

4.3   Algorithm of the program

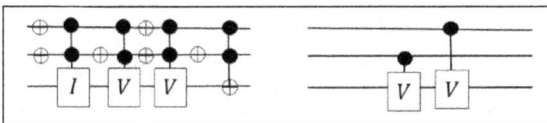

Figure 13: A standard form binary multiplexer (left) and its FPQF equivalent in polarity 11 (right). Note the drastic simplification of the circuit.

A program was written to automate the process of generating KQF butterflies and calculating costs of the resulting circuits. The program is written in Java because of its strong Object Oriented Programming (OOP) support; the code uses many classes (such as Multiplexer, Function, and more), and thus, JavaâĂŹs strong support of OOP is a large benefit. When executed, the program asks for the target functions of the standard form multiplexer in the form of unitary matrices as input (which are inputted in ascending natural order based off the index of the controls), and outputs the least expensive polarity of FPQF.

Alternatively, the program can also parse the names of common functions (e.g. âĂIJPXâĂİ for Pauli X, or âĂIJHâĂİ for Hadamard) to unitary matrices. The program uses a Depth First Search algorithm to generate all possible KQF forms by recursively applying layers of butterflies; that is, for each layer of butterflies,

21

the program first applies a layer of polarity 0 butterflies to the multiplexer and recursively calls itself; after the call returns, the program backtracks by one layer and applies a layer of polarity 1 butterflies to the multiplexer, then recursively calls itself, etc. Butterflies are implemented as a series of matrix transformations, as seen in Figure 11. The polarity 0 butterfly receives three unitary matrices, a, b, and c. It then outputs c as the first output, $a \cdot c^{-1}$ as the second output, and $b \cdot c^{-1}$ as the third output. These outputs can be encoded as the combinations of matrix inversions and matrix multiplications. The other butterfly polarities can be coded similarly. Once all KQF forms are generated, cost can be calculated as previously discussed: for each controlled function, determine the number of controls n needed for that function. If $n < 10$, the program references the costs derived by [8]. Otherwise, the code uses the equation $32m - 96$ to determine the cost of the gate.

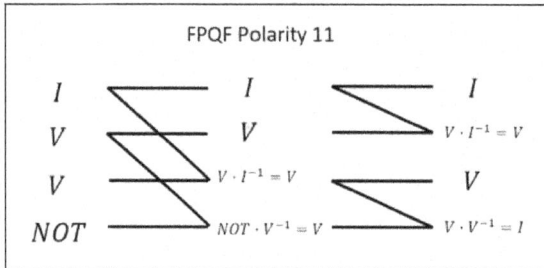

Figure 14: The butterfly diagram used to convert the standard form multiplexer to FPQF polarity 11 in Figure 13.

By finding the costs of all KQF polarities, the program can exhaustively determine which polarities are the cheapest, thus finding the exact minimum cost for an FPQF or KQF realization of a multiplexer.

5   Analysis and results

An additional program was written (also in Java) to generate large random test cases; initially, the test multiplexers were created by randomly pulling target functions from a set of common quantum functions: Pauli Rotations, Hadamard, NOT, V, V Hermitian, and Identity. A set of well-known quantum circuit cases were also created manually, such as a binary case that involves Identity, V, V, and NOT, as seen in Figure 13. The corresponding butterfly diagram for the case in Figure 13 can be found in Figure 14. Another example is provided in Figure 15 of a standard form ternary multiplexer and its drastically cheaper FPQF counterpart. On these well-

Number of Controls	Original Cost	Best Polarity	Worst Polarity	Average Polarity	Best Cost Reduction
3	322	148	322	210	54%
4	2100	1007	2100	1276	52%
5	11713	5248	11713	6727	55%
6	52785	24722	52785	31798	53%
7	219258	108331	219258	138359	51%
8	874045	442109	874045	562295	49%
8	869705	442573	869705	562481	49%
8	870170	442521	870170	562351	49%

Table 2: Data on ternary cases generated with Paulis, Hadamards, NOTs, V/V Hermitians, and Identities. The program generated all KQF polarities.

known cases, the program correctly generated all polarities, including the optimal solutions.

5.1 Randomly generated cases: Pauli X/Y/Z, Hadamard and H Hermitian, NOT, V and V Hermitian, Identity

Cases with more variables take exponentially longer time to run, so it became impractical to run the program on test cases with more than 9 control variables. In the best case, the KQF forms that were generated had costs as little as 50% of the cost of the standard form multiplexer, and the cost difference between the average polarity and the optimal polarity was significant: see Table 2 for results on several examples. Note that the best case cost reduction decreases slightly as the size of the multiplexer increases, which suggests that the method becomes less effective as multiplexer size increases.

It is interesting to note that the optimal polarities (not shown in the tables) were all FPQF forms; Shannon expansions were not a part of the optimal solution for any of the test cases. This suggests that for this type of randomly generated data, having a Shannon expansion does not improve the cost of the result.

5.2 Randomly generated cases: NOT, V and V Hermitian

This method was also tested on quantum multiplexers randomly generated from a much smaller pool of target gates: NOT, V, and V Hermitian only. We are interested in using this set of gates as targets because there are several algorithms to synthesize quantum reversible circuits with this set of gates [8]. In addition, this

Number of Controls	Original Cost	Best Polarity	Worst Polarity	Average Polarity	Best Cost Reduction
3	378	133	378	180	65%
4	2430	737	2430	1029	70%
5	12879	3710	12879	5210	71%
6	61965	17260	61965	24438	72%
7	255879	77828	255879	105731	70%
8	1016955	324077	1016955	428118	68%
8	1016955	325687	1016955	427870	68%
8	1016955	326620	1016955	427705	68%
9	3798819	1277506	3798819	1649713	66%
9	3798819	1280032	3798819	1649279	66%

Table 3: Data on ternary cases generated with NOTs, Vs, and V Hermitians. The program generated all KQF polarities.

set of gates is very conducive for canceling to Identity, so we expect best case cost reduction to be even greater for this set of data. It can be seen in Table 3 that the cost reduction is significantly greater (at 60-70%), and as before, the significant cost difference between the best case and average case costs means that it is meaningful to search for the best polarity instead of only using a random one. Once again, the effectiveness of the method decreased as size increased. Note once more that randomly generated functions are the most difficult cases.

Once more, the KQF method did not offer improvements on the best case cost; all of the optimal butterflies were FPQF forms.

6   Conclusion

In this paper, we first define standard form multiplexers and two new types of multiplexer forms: FPQF and KQF forms. Next, we provide a method to convert a standard form multiplexer to an FPQF or KQF form using a polarity transformation, and we give a proof for decomposing these transformations into smaller transformations that can be represented with butterfly diagrams in an analogous way to well-known FPRM and KRM butterfly diagrams. Note that unlike work on ESOP minimization [6, 11, 12, 19], our work is an extension of Reed Muller to quantum circuits where the target gate can be any arbitrary target function instead of only NOT, like in classical Reed Muller. Then, we test this method for randomly generated standard form ternary multiplexers that use target functions from a small pool of common quantum functions, finding that cost reductions are as high as 70%, but decrease as size (and equivalently, complexity) of the standard multiplexer increases. We also find that there are significant cost differences between the best case polarity and the average case polarity, which justify the long runtime required to exhaustively apply Depth First Search to find the maximally optimized KQF form. Additionally, we discovered that on randomly generated data, KQF offers no benefit on best case cost reduction (the best case answers are always FPQF forms) while having a longer run time than a program that searched among FPQF forms only. This suggests that on randomly generated multiplexers, there is no benefit to using KQF; only FPQF should be used to optimize these types of multiplexers.

The FPQF and KQF methods can also be extended to any multivalued quantum multiplexers, not just ternary or binary: it is expected that the effectiveness and cost reductions of these methods will be very high for multivalued logics with $q > 3$ as well. However, the time costs of the current algorithm are exponentially great due to the use of complete search, which suggests low scalability potential for circuits with high variable counts. Furthermore, the steadily decreasing trend of effectiveness of this method on larger randomly generated multiplexers suggests that more research should be conducted to find alternate methods for optimizing larger standard form multiplexers since it appears that the FPQF

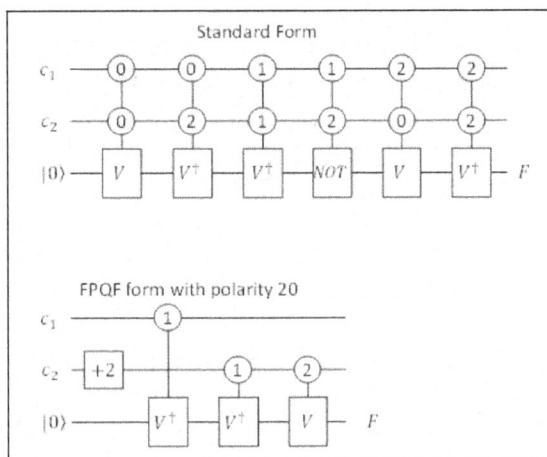

Figure 15: A standard form ternary multiplexer and its FPQF equivalent in polarity 20.

method will not net many savings at extremely high sizes. This is consistent with findings in [11] and [12], which suggest that Reed Muller-based methods, such as FPRM, KRM, and even GRM [12], lead to non-minimal results when minimizing Boolean functions that contain minterms with high Hamming distances. Finally, all our methods can be extended to incompletely specified functions, generalizing the method from [16].

References

[1] R. R. Tucci, "A Rudimentary Quantum Compiler (2cnd Ed.)," arXiv, 2008.

[2] J. J. Vartiainen, M. Mãűttãűnen and M. M. Salomaa, "Efficient decomposition of quantum gates," *Physical Review Letters*, vol. 92, no. 17, p. 177902, 2004.

[3] V. V. Shende, S. S. Bullock and I. L. Markov, "Synthesis of Quantum Logic Circuits," *IEEE Transactions on Computer-Aided Design of Integrated Circuits and Systems*, vol. 25, no. 6, pp. 1000-1010, 2006.

[4] V. V. Shende, I. L. Markov and S. S. Bullock, "Minimal Universal Two-Qubit CNOT-based Circuits," *Physical Review A*, vol. 69, no. 6, p. 062321, 2004.

[5] M. Davio, J.-P. Deschamps and A. Thayse, Discrete and Switching Functions, St-Saphorin: Georgi Publishing Company, 1978.

[6] M. Perkowski and D. Shah, Design of Regular Reversible Quantum Circuits, Portland: Portland State University, 2010.

[7] D. Maslov and G. W. Dueck, "Improved Quantum Cost for n-bit Toffoli Gates," *Electronic Letters*, vol. 39, no. 25, pp. 1790-1791, 2004.

[8] A. Barenco, C. H. Bennett, R. Cleve, D. P. DiVincenzo, N. Margolus, P. Shor, T. Sleator, J. Smolin and H. Weinfurter, "Elementary gates for quantum computation," *Physical Review A*, vol. 52, no. 5, pp. 3457-3467, 1995.

[9] S. R. Hutsell, "An Eigenanalysis and Synthesis of Unitary Operators used in Quantum Computing Algorithms," Portland State University, PHD dissertation, Portland, 2009.

[10] D. H. Green, "Families of Reed-Muller canonical forms," *International Journal of Electronics*, vol. 70, no. 2, pp. 259-280, 1991.

[11] A. Mishchenko and M. Perkowski, "Fast Heuristic Minimization of Exclusive-Sums-of-Products," *5th International Reed-Muller Workshop*, 2001.

[12] L. Csanky, M. A. Perkowski and I. Schafer, "Canonical restricted mixed-polarity exclusive-OR sums of products and the efficient algorithm for their minimisation," *IEE Proceedings E - Computers and Digital Techniques*, vol. 140, no. 1, pp. 69-77, 1993.

[13] C. Gidney, "Halving the cost of quantum addition," *Quantum*, vol. 2, p. 74, 2018.

[14] M. Soeken, M. Roetteler, N. Wiebe and G. D. Micheli, "Hierarchical Reversible Logic Synthesis Using LUTs," in *54th Annual Design Automation Conference*, Austin, 2017.

[15] G. Meuli, M. Soeken, M. Roetteler, N. Wiebe and G. D. Micheli, "A best-fit mapping algorithm to facilitate ESOP-decomposition in clifford+T quantum network synthesis," in *23rd Asia and South Pacific Design Automation Conference*, Jeju, 2018.

[16] D. Debnath and T. Sasao, "Exact Minimization of FPRMs for Incompletely Specified Functions by Using MTBDDs," *IEICE Transactions*, Vols. E88-A, no. 12, pp. 3332-3341, 2005.

[17] W. N. N. Hung, G. Yang, X. Song and J. Yang, "Optimal synthesis of multiple output Boolean functions using a set of quantum gates by symbolic reachability analysis," *IEEE Transactions on Computer-Aided Design of Integrated Circuits and Systems*, pp. 1652-1663, 2006.

[18] H. Wu, M. A. Perkowski, X. Zeng and N. Zhuang, "Generalized partially-mixed-polarity Reed-Muller expansion and its fast computation," *IEEE Transactions on Computers*, vol. 45, no. 9, pp. 1084-1088, 1996.

[19] R. Drechsler, A. Finder and R. Wille, "Improving ESOP-Based Synthesis of Reversible Logic Using Evolutionary Algorithms," in *Applications of Evolutionary Computation*, Torino, 2011.

[20] E. Tsai and M. Perkowski, "Synthesis of Permutative Quantum Circuits with Toffoli and TISC Gates," in *The International Symposium on Multiple-Valued Logic*, Tuusula, 2012.

[21] F. S. Khan, Quantum Multiplexers, Parrondo Games, and Proper Quantization, Portland: Portland State University, 2009.

[22] F. S. Khan and M. Perkowski, "Synthesis of multi-qudit Hybrid and d-valued Quantum Logic Circuits by Decomposition," *Theoretical Computer Science*, vol. 367, no. 3, pp. 336-346, 2006.

[23] S. A. Bleiler and F. S. Khan, "Properly quantized history-dependent Parrondo games, Markov processes, and multiplexing circuits," *Physics Letters A*, vol. 375, no. 19, pp. 1930-1943, 2011.

[24] M. H. A. Khan, "Primitive quantum gate realizations of multiple-controlled Toffoli gates," in *16th Int'l Conf. Computer and Information Technology*, Khulna, 2014.

[25] Z. Li, M. Perkowski, X. Song and H.-W. Chen, "Realization of a new permutative gate library using controlled-kth-root-of-NOT quantum gates for exact minimization of quantum circuits," *International Journal of Quantum Information*, 2014.

[26] Z. Li, X. Song, S. Chen and M. Perkowski, "Quantum Circuit Synthesis using a New Quantum Logic Gate Library of NCV Quantum Gates," *International Journal of Theoretical Physics*, 2016.

[27] M. Lukac and M. Perkowski, "Quantum Behaviors: Synthesis and Measurement," in *RM 2007*, Oslo, 2007.

[28] Y. Wang and M. Perkowski, "Improved Complexity of Quantum Oracles for Ternary Grover Algorithm for Graph Coloring," in *IEEE 41st International Symposium on Multiple-Valued Logic*, Tuusula, 2011.

[29] K. Jin, T. Saffat and M. Perkowski, "A fixed polarity approach for optimization of general binary quantum multiplexers," in *International Workshop on Quantum Compilation*, San Diego, 2018.

Received 30 April 2019

PAM-4 SIGNAL TRANSMITTER USING FPGA AND DAC FOR SERIAL-LINK TEST

YASUSHI YUMINAKA
Graduate School of Science and Technology, Gunma University, JAPAN
yuminaka@gunma-u.ac.jp

NATSUKI SATO
Graduate School of Science and Technology, Gunma University, JAPAN

TAKAHITO CHIGIRA
Graduate School of Science and Technology, Gunma University, JAPAN

KOHEI TOYODA
School of Science and Technology, Gunma University, JAPAN

YOSUKE IIJIMA
National Institute of Technology (KOSEN), Oyama College, JAPAN

Abstract

As next-generation high-speed data transmission standards, 4-level pulse amplitude modulation (PAM-4) signaling is adopted to support 400 Gbps data transmission. This paper proposes a multi-valued signal transmitter for PAM-4 serial-link tests. An arbitrary signal generator was designed using a high-speed digital-to-analog converter (DAC) and an FPGA. This system can be used for evaluating feed-forward equalization (FFE) and predistortion at a transmitter side by emulating a transmitter for PAM-4 serial link. FFE and predistortion design examples for PAM-4 signaling are shown and experimentally verified using the FPGA-based DAC.

This paper is an extension of [5]

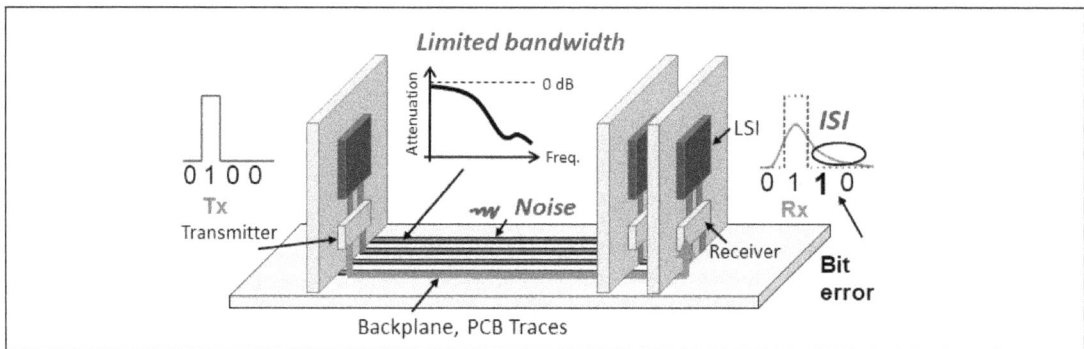

Figure 1: Problems in high-speed serial links.

1 Introduction

There is a growing demand for ultra-high-speed serial electrical interconnects beyond several tens of Gbps (Gigabit per second) for chip-to-chip links, backplanes, and data centers, as well throughout the IT infrastructure. However, signal distortions due to, e.g., inter-symbol interference (ISI) caused by the limited bandwidth of the channel and noise significantly limit the I/O bandwidth, and thereby restricting the total VLSI system performance (Fig.1). Conventional binary non-return-to-zero (NRZ) signal can no longer economically maintain the signal integrity required for reliable data transfer. Alternatively, 4-level pulse amplitude modulation (PAM-4) proves effective to mitigate bandwidth limitations. PAM-4 represents two consecutive binary NRZ bits (**00,01,10,11**) using one symbol (**0,1,2,3**) at a two-times slower symbol rate. This can halve the channel bandwidth limitation compared with NRZ signaling.

In accordance with next-generation high-speed data transmission standards IEEE 802.3bs [1] and CEI-56G [2], PAM-4 signaling is employed as a promising approach to support 400 Gbps data transmission (8 lanes of 25 Gbaud with PAM-4 signaling). In response to the trends, the equipment and software applications for PAM-4 test solutions are provided by test and measurement suppliers [3]. However, the measuring apparatus and software are very expensive and lack of flexibility for modifying the function in accordance with future testing methods [4].

To overcome the aforementioned problems, this paper proposes a multi-valued transmitter using a high-speed digital-to-analog converter (DAC) and an FPGA for serial-link tests [5]. Using the FPGA and DAC allows for sophisticated digital equalization, e.g. feed-forward equalizer (FFE) and supports a digital predistortion scheme. Simulation and actual measurement results including equalization and predistortion techniques are shown to demonstrate the feasibility of the PAM-4

30

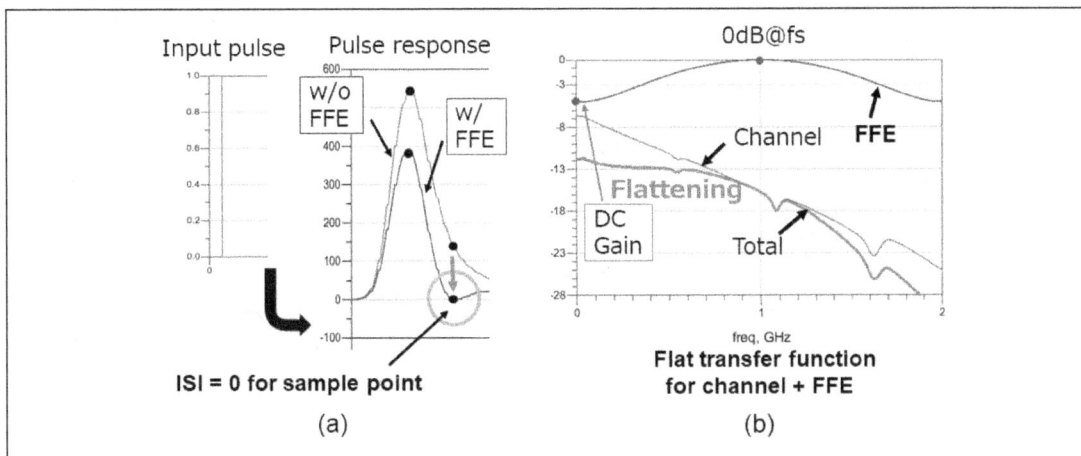

Figure 2: FFE design approaches. (a) Time domain, (b) frequency domain.

serial-link test systems.

This paper is organized as follows: Section 2 discusses PAM-4 signaling and FFE design, and FPGA and DAC implementation for the PAM-4 transmitter. Section 3 presents the simulation and measurement results of equalization and predistortion techniques. Finally, Section 4 concludes the paper.

2 Transmitter for PAM-4 Signaling

2.1 PAM-4 Signaling and Feed-Forward Equalizer Design

PAM-4 coding can represent 2 bit information as one symbol using a 4-level signal, resulting in reducing interconnection by half. Furthermore, the use of PAM-4 signaling can reduce the symbol rate by a factor of 2 compared to NRZ binary signaling, and thus, lowers the bandwidth requirements of the channel. It is possible to relax not only data rate but also operating frequency of transmitter and receiver circuitry [6]-[9].

In addition to the PAM-4 efficient coding scheme, signal processing techniques are required for compensating for the deterioration of the signal in high-speed serial-link [10]. For instance, FFE at the transmitter is an effective technique that boosts the high-frequency components of the deteriorated signal to invert the low-pass characteristic of the channel. The purpose of channel equalization is to flatten the response within the frequency of interest in the frequency domain to remove pre- and post-cursor ISI in the time domain.

For example, the 3-tap FFE example is constructed with the sum of the product

31

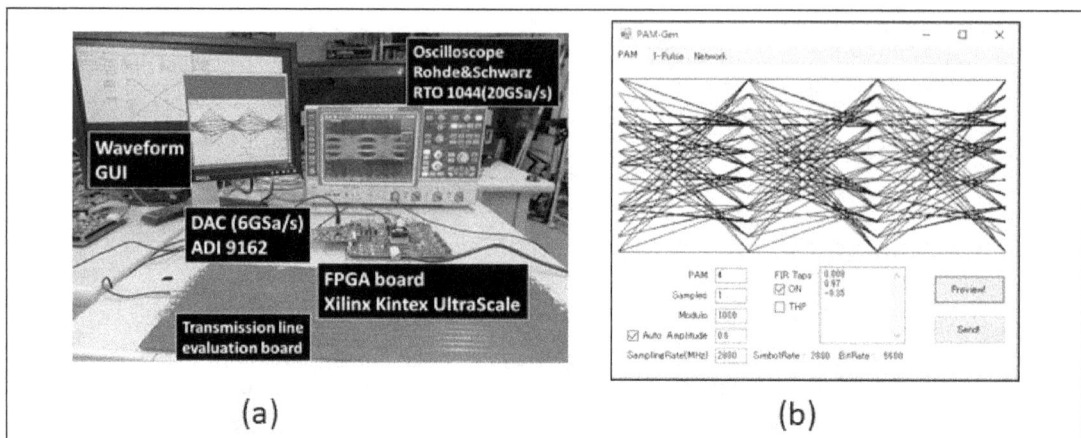

Figure 3: Serial-link evaluation system. (a) System setup (Xilinx Kintex-7 FPGA and Analog Devices AD9739A DAC), (b) GUI for transmitter, which can set, e.g., PAM level, symbol rate, and equalization coefficient setting.

of coefficients and input signals with delay. The tap coefficients can be derived from the unit pulse response of the channel to realizes ISI=0 for sample point of pulse response by using matrix operation and optimization techniques. As shown in Fig.2(a), the deteriorated pulse X is expressed by the ideal pulse matrix Z and FFE coefficients matrix C as $Z = CX$. Then, FFE coefficients matrix can be obtained as $C = ZX^{-1}$ based on least mean square error or zero-forcing techniques [11].

In contrast to above time-domain design procedure, more intuitive and simple way is the frequency-domain design approach that targets to flatten the total transfer function of the channel and FFE. Design procedure of 3-tap FFE is as follows (Fig.2(b)).

(1) Channel characteristics are given (measured by a network analyzer).

(2) We define the transfer function of the FFE as the equation $H(z) = c_{-1} + c_0 z^{-1} + c_1 z^{-2}$.

(3) The magnitude of FFE becomes 0dB at the sampling frequency, which is same as the Nyquist frequency. We set the slope of the FFE response by changing the DC gain ($20 \log(c_{-1}+c_0+c_1)$) of the FFE that can flatten the total response between the DC and Nyquist frequency of signals.

32

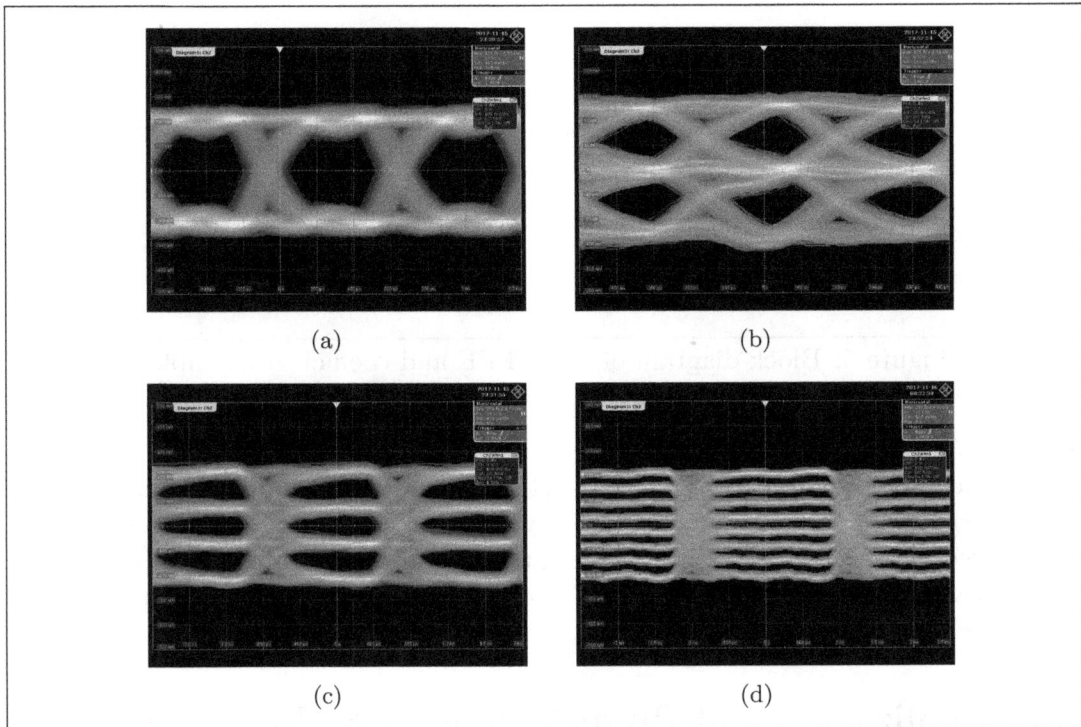

Figure 4: Modulation waveform examples of the transmitter based on FPGA-based DAC (1.4 Gsps). (a) Binary (NRZ), (b) Duobinary, (c) PAM-4, (d) PAM-8.

2.2   Transmitter Design Using FPGA and DAC

In this subsection, we propose an arbitrary signal generator for a PAM-4 serial-link test using a high-speed DAC and an FPGA. A high-speed FPGA Kintex UltraScale is used to control the DAC (Analog devices, AD9162; 6 GSa/s), as shown in Fig.3(a). To establish a link between data converters (ADCs and DACs) and FPGA at beyond 3 Gbps, legacy electrical bus such as LVDS is not sufficient. JESD204B is standard [12] which makes it possible to achieve serial interface up to 12.5 Gbps. Because JESD204B IP core for FPGA is expensive, we originally developed HDL code for JESD204B according to the JESD204B specification.

Moreover, an arbitrary waveform generator GUI using the FPGA-based DAC was developed that can set, e.g., the PAM level and symbol rate (Fig.3(b)). The GUI supports several modulation formats such as NRZ, Duobinary, PAM-4, and PAM-8, as shown in Figs.4(a)-(d). Furthermore, GUI can emulate a transmitter equalization function such as FFE, Tomlinson-Harashima precoding with modulo

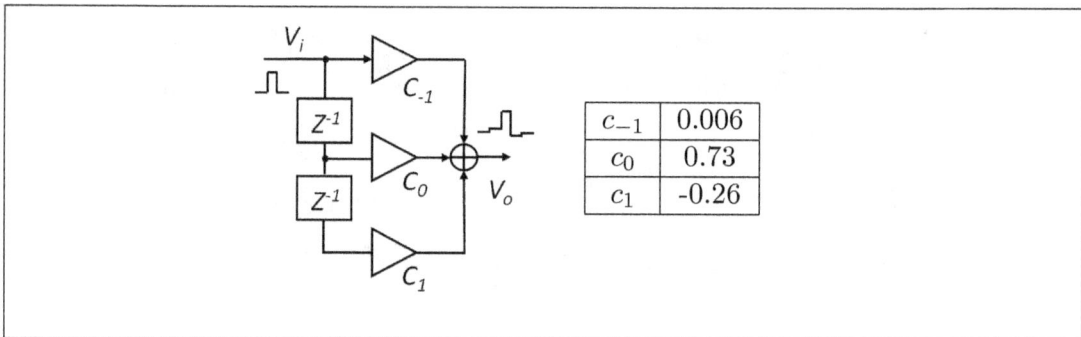

c_{-1}	0.006
c_0	0.73
c_1	-0.26

Figure 5: Block diagram of 3-tap FFE and coefficient example.

operation [6]-[9].

In the proof-of-concept prototype system, we used peripheral circuits on the FPGA evaluation board, such as memory and interface chips. Coefficients and modulated data calculated by GUI are sent to the memory and then transferred to an FPGA to control a high-speed DAC for generating coding and FFE modulated signaling.

3 Equalization and Predistortion of PAM-4 Signal

In this section, we discuss equalization and predistortion techniques to demonstrate the feasibility of the PAM-4 transmitter based on the application of an FPGA-based DAC.

3.1 Equalization for PAM-4 Using FFE

Transmitter FFE, and passive continuous-time linear equalization (CTLE) and decision feedback equalizer (DFE) used in the receiver are effective techniques that boost the high-frequency components of the waveform relative to its low-frequency components to invert the effect of the channel. As shown in Fig.5, our system supports transmitter FFE by setting coefficients using an FPGA. Under the signal and channel conditions (2.8 Gsps PAM-4, micro-strip line of 1 m with a channel loss of -13.5 dB at 2.8 GHz), the transfer function of the 3-tap FFE becomes

$$H(z) = 0.006 + 0.73z^{-1} - 0.26z^{-2}, \qquad (1)$$

which has a DC gain of $20\log(0.006 + 0.73 - 0.26)$ = -6.55 dB and 0 dB at the sampling frequency. These coefficients were calculated by the FFE design procedure described in Section 2.

This high-pass filter response of FFE compensated for the low-pass characteristics of the channel to flatten the magnitude response for the signal bandwidth. The simulation and measured results of FFE using an FPGA-based DAC are shown in Figs.6(a)-(d). As shown in these eye diagrams, the PAM-4 eye is opened by FFE, thereby improving signal integrity. Using the FPGA, the arbitrary FFE digital filter is realized and the FFE test is possible.

Moreover, the GUI (Fig.3(b)) supports Tomlinson-Harashima precoding with modulo operation [7],[8]. Advanced CMOS technology scaling enables us to implement flexible on-chip digital signal processing algorithms for equalization and a more bandwidth-efficient modulation scheme. This motivates the use of an FPGA and a DAC-based transmitter design. The digital-rich transmitter architecture allows for resilience against process variations regarding fine CMOS technology nodes compared with conventional analog implementation.

3.2 Predistortion for PAM-4 Signal

As an another application of FPGA-based DAC, we demonstrate the predistortion techniques. In the high-frequency operation, nonideal amplifiers of drivers at a transmitter cannot keep a constant gain with increasing input signal. Namely, the gain of logic levels 2 and 3 will be below that of logic level 1 when the PAM-4 input signal is too large. This will distort the PAM-4 eye diagrams.

The transmitter linearity, which indicates a vertical linearity, is defined as the level separation mismatch ratio R_{LM} [3]. The effective symbol levels ES_1 and ES_2 are the average levels of the two center symbols with

$$ES_1 = (V_1 - V_{mid})/(V_0 - V_{mid}) \qquad (2)$$
$$ES_2 = (V_2 - V_{mid})/(V_3 - V_{mid}). \qquad (3)$$

The level separation mismatch ratio R_{LM} is defined as

$$R_{LM} = min((3 \cdot ES_1), (3 \cdot ES_2), (2 - 3 \cdot ES_1), (2 - 3 \cdot ES_2)),$$

$$(4)$$

where V_{mid} is the average voltage of the four levels defined as $V_{mid} = \frac{1}{2}(V_0 + V_3)$, as shown in Fig.7. An ideal linear transmitter yields the symbol separations $ES_1 = ES_2 = 1/3$ and $R_{LM} = 1$.

The amplitude distortion is caused by the compression/expansion nonlinearities in the Tx amplifiers, as illustrated in Fig. 8 (Agilent INA-02186: 31 dB gain). The included nonlinearity causes an unbalanced vertical eye opening, thereby resulting

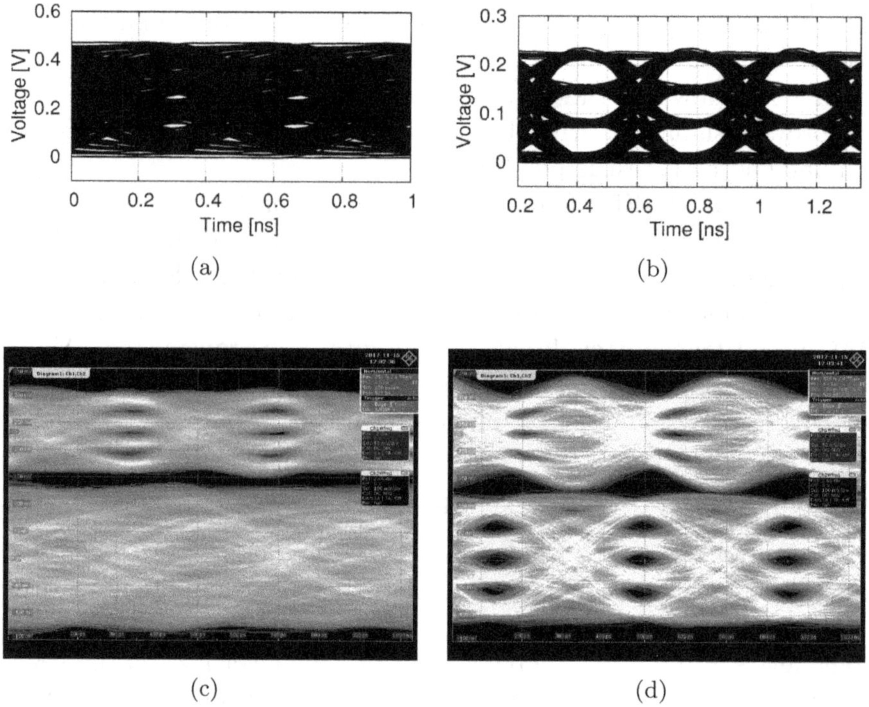

Figure 6: FFE using FPGA-based DAC (2.8 Gsps PAM-4, micro-strip line of 1 m with channel loss of -13.5 dB at 2.8GHz). (a) Simulated eye diagram at Rx w/o FFE, (b) simulated eye diagram at Rx w/ FFE, (c) measured eye diagram at Tx and Rx w/o FFE, (d) measured eye diagram at Tx and Rx w/ FFE.

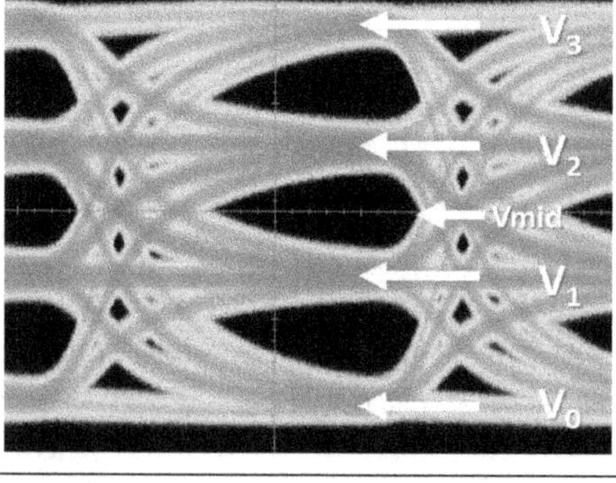

Figure 7: Transmitter linearity test based on R_{LM}.

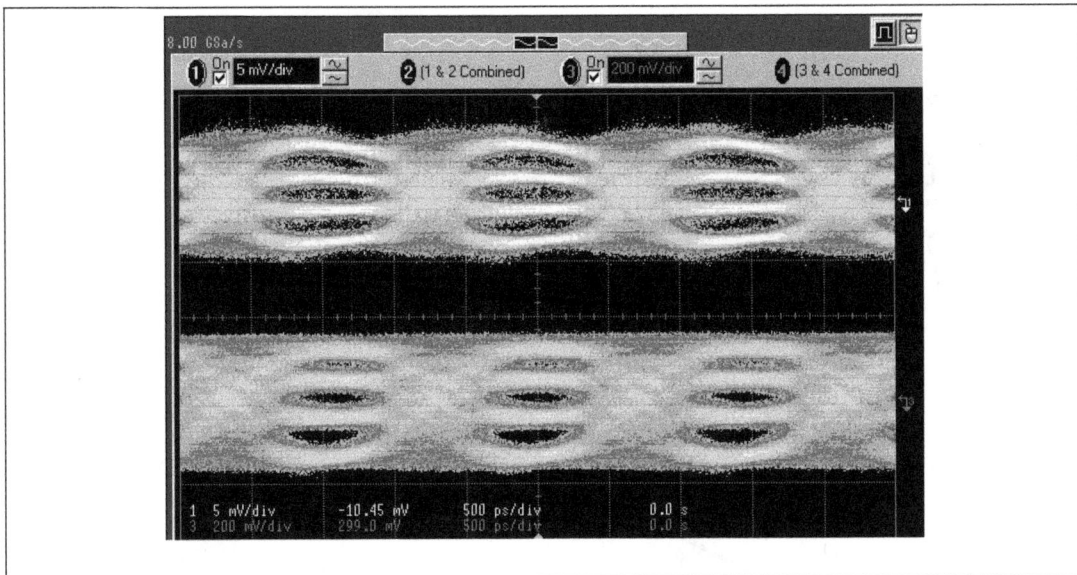

Figure 8: PAM-4 Eye diagram with nonlinearity. DAC output of PAM-4 signal (top) and amplifier output (bottom).

in a worse R_{LM} (top: DAC output; bottom: amplifier output). In this case, the amplitude level spacings between level 3 and 2 and between 2 and 1 are compressed compared to those for 1 and 0.

Tx data amplitude predistortion techniques are employed to address the amplitude distortion, as shown in Fig. 9. The Tx amplifier exhibits a nonlinear relation between the logic level and the actual amplitude level with **0**: 0 V, **1**: 0.37 V, **2**: 0.7 V, and **3**: 1 V as shown in Fig.9(a) (Ideally, **0**: 0 V, **1**: 0.33 V, **2**: 0.67 V, and **3**: 1 V). To linearize the relation (Fig.9(c)), we preassign the amplitude level to the logic value. This is accomplished with predistortion of the voltage assignment versus each logic value using an FPGA-controlled DAC. In this case, the assignment is **0**: 0 V, **1**: 0.3 V, **2**: 0.64 V, and **3**: 1 V as shown in Fig.9(b). Consequently, it is possible to ensure that the vertical eye opening has equally spaced levels.

Figures 10(a) and 10(b) show the measured eye diagrams with and without predistortion. Before predistortion, the eye diagram exhibits $ES_1 = 0.25, ES_2 = 0.42$, and $R_{LM} = 0.74$. By applying predistortion, these parameters are improved to $ES_1 = 0.37, ES_2 = 0.33$ and $R_{LM} = 0.90$ as listed in Table 1.

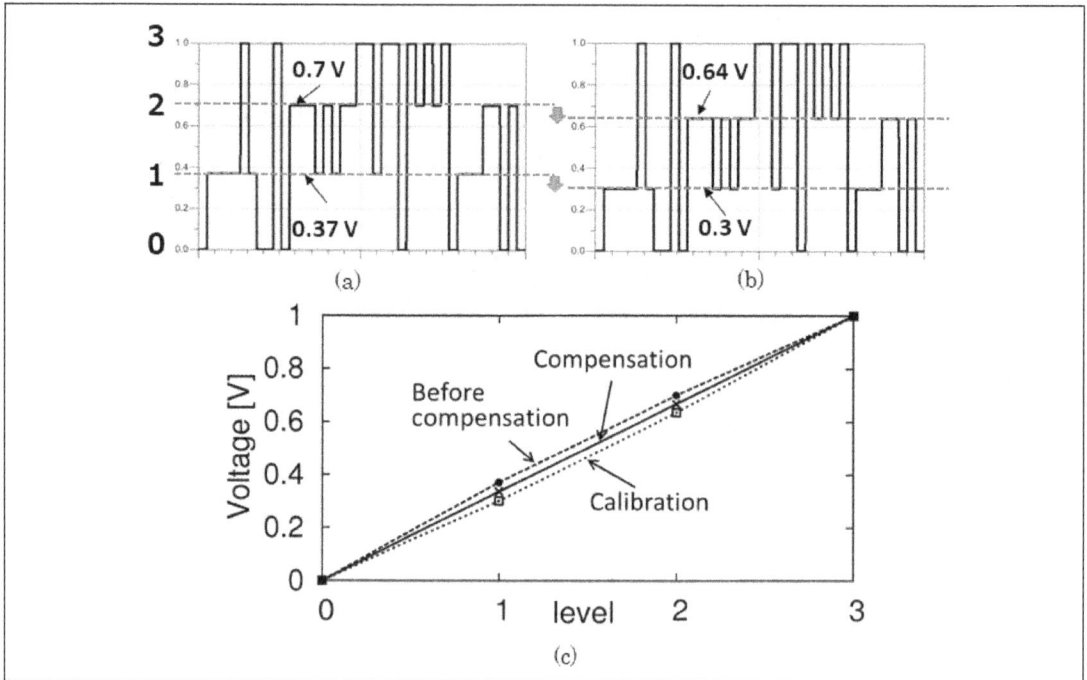

Figure 9: Predistortion technique: (a) logic value and voltage at a nonlinear amplifier output, (b) logic value and voltage assignment using predistortion, (c) relation between logic value and voltage.

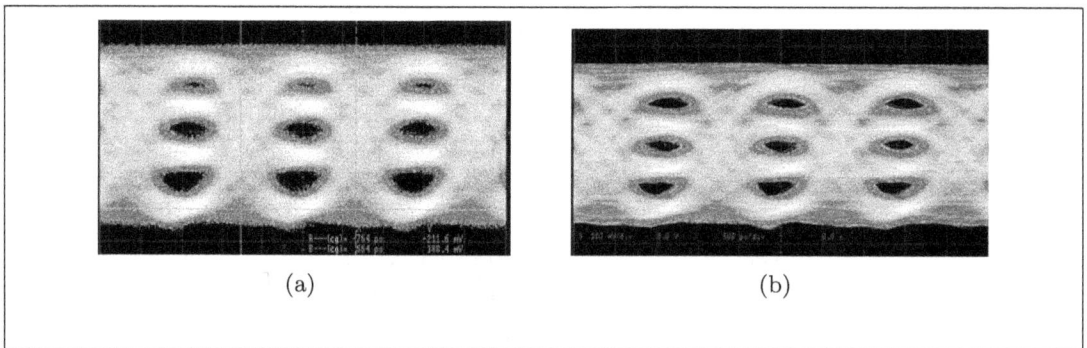

Figure 10: PAM-4 Eye diagram with nonlinearity and its compensation using predistortion. (a) measured eye diagram w/o predistortion, (b) measured eye diagram w/ predistortion.

	ES_1 [V]	ES_2 [V]	R_{LM}
Without compensation	0.25	0.42	0.74
With compensation	0.37	0.33	0.90

Table 1: Evaluation of nonlinearity compensation using predistortion.

4 Conclusion

The use of PAM-4 in inter-/intra-chip communication demands sophisticated test and measurement techniques for interface interoperability and test validation. This paper proposed an arbitrary signal generator using a high-speed DAC and an FPGA for PAM-4 serial-link tests. The proposed system can be used for evaluating FFE and predistortion at the transmitter side by emulating a transmitter for PAM-4 serial link. Future work includes a realization of more sophisticated equalization algorithm and adaptive coefficient calibration for FFE.

Acknowledgment

This work was supported by JSPS KAKENHI Grant Number JP101488.

References

[1] IEEE P802.3bs 200 Gb/s and 400 Gb/s Ethernet Task Force, [Online] Available: http://www.ieee802.org/3/bs/

[2] OIF CEI-56G Application Note. [Online] Available: http://www.oiforum.com/wp-content/uploads/OIF-CEI-white-paperfinal-Mar-23-2016.pdf

[3] PAM4 Signaling in High Speed Serial Technology: Test, Analysis, and Debug, Tektronix Application Note.

[4] J. Moreira, H. Werkmann, M. Ishida, B. Roth, V. Filsinger and S-.X. Yang, "An ATE Based 32 Gbaud PAM-4 At-Speed Characterization and Testing Solution," *Proc. 23rd Asian Test Symposium*, pp.218-223, 2014.

[5] N. Sato, T. Chigira, K. Toyoda, Y. Iijima and Y. Yuminaka, "Multi-Valued Signal Generation and Measurement for PAM-4 Serial-Link Test," *Proc. IEEE 48th International Symposium on Multiple-Valued Logic*, pp.210-214, 2018.

[6] Y. Yuminaka, Y. Takada and T. Okada, "Comparison of Spectrally Efficient Coding Techniques for High-Speed VLSI Data Transmission," *Journal of Multiple-Valued Logic and Soft Computing*, Vol.26, pp.75-87, 2016.

[7] Y. Yuminaka and Y. Iijima, "Multiple-Valued Signaling for High-Speed Serial Links Using Tomlinson-Harashima Precoding," *IEEE Journal of Emerging and Selected Topics in Circuits and Systems*, Vol.6, No.1, pp.25-33, 2016.

[8] Y. Iijima and Y. Yuminaka, "Double-Rate Tomlinson-Harashima Precoding for Multi-Valued Data Transmission," *IEICE Transactions on Information & Systems*, Vol.100-D, 8, pp.1611-1617, 2017.

[9] Y. Yuminaka, T. Kitamura and Y. Iijima, "PAM-4 Eye Diagram Analysis and Its Monitoring Technique for Adaptive Pre-Emphasis for Multi-valued Data Transmissions," *Proc. IEEE 47th International Symposium on Multiple-Valued Logic*, pp.13-18, 2017.

[10] M. Bassi, F. Radice and M. Bruccoleri, "A High-Swing 45 Gb/s Hybrid Voltage and Current-Mode PAM-4 Transmitter in 28 nm CMOS FDSOI," *IEEE Journal of Solid-State Circuits*, Vol.51, No. 11, pp.2702-2715, 2016.

[11] S. Hall and H. Heck: *Advanced Signal Integrity for High-Speed Digital Designs*, Wiley 2009.

[12] JESD204 High Speed Interface, [Online] Available: https://www.xilinx.com/products/technology/high-speed-serial/jesd204-high-speed-interface.html

Received 30 April 2019

Fast hardware-based learning algorithm for binarized perceptrons using CMOS invertible logic

Naoya Onizawa

Research Institute of Electrical Communication, Tohoku University, Japan
naoya.onizawa.a7@tohoku.ac.jp

Duckgyu Shin

Research Institute of Electrical Communication, Tohoku University, Japan
duckgyu.shin.p4@dc.tohoku.ac.jp

Takahiro Hanyu

Research Institute of Electrical Communication, Tohoku University, Japan
hanyu@riec.tohoku.ac.jp

Abstract

This paper introduces a fast hardware-based learning algorithm for perceptrons using CMOS invertible logic. CMOS invertible logic is designed based on underlying Boltzmann machines that probabilistically realizes forward and backward operations using stochastic computing. This bidirectional-computing capability enables us to directly obtain weights of the perceptron without calculating a loss function used in a traditional learning algorithm. As a result, the proposed invertible-learning algorithm can perform with parallel training data as opposed to a sequential learning process of the traditional algorithm. For performance evaluation, a 25-input binarized perceptron is learned using a simplified Modified National Institute of Standards and Technology (MNIST) dataset. The proposed learning speed estimated using a 65-nm CMOS technology can be around a 5,600 x faster than the traditional perceptron-based learning algorithm, while maintaining a similar accuracy of 98%.

1 Introduction

Invertible logic has been recently presented for providing a capability of probabilistically forward and backward operations [1, 2] as opposed to typical binary logic for the forward operation. It is designed based on underlying Boltzmann machines [3] and probabilistic magnetoresistive device models (p-bits) [4] whose input and output signals are represented by random bit streams. Differing from invertible logic is reversible logic [5, 6], where circuits are constructed of special gates (such as Controlled NOT or Toffoli gates) having a direct one-to-one mapping of inputs to outputs with deterministic computing (see details in [7]). The bidirectional-computing capability is realized by reducing the network energies with noise control (e.g. a multiplier could be used as a factorizer in the backward mode or a divider in the partial forward/backward mode). Several challenging problems could be quickly solved using invertible-logic hardware, such as integer factorization (e.g. cryptography problems [8]) and combinational optimization (e.g. wireless sensor networks [9]).

However, as invertible logic in [1] is designed using the device model, it is difficult to implement the invertible-logic hardware. The device model can be approximated using digital circuits with binary logic [10] or stochastic computing [7]. These approximation techniques enable us to implement the invertible-logic hardware in field-programmable gate arrays (FPGAs) and application-specific integrated circuits (ASICs). Currently, the hardware implementation of invertible logic is limited to invertible adders (subtracters) and invertible multipliers (factorizers/dividers).

In this paper, we expand the application of invertible logic to a perceptron [11] for the future learning hardware of neural networks. The perceptron is a single-layer neural networks, while a multi-layer perceptron is a class of feed-forwarded neural networks. An invertible perceptron with a binary precision is designed based on CMOS invertible logic with stochastic computing [7]. Stochastic computing [12, 13] realizes area-efficient arithmetic circuits using random bit streams. The proposed invertible binarized perceptron directly obtains weights using training data and labels without calculating a loss function used in a traditional perceptron-based learning algorithm. This hardware-based learning can be realized in parallel as opposed to a sequential process of the traditional learning algorithm in software. Using a simplified Modified National Institute of Standards and Technology (MNIST) dataset [14], the proposed 25-input invertible binarized perceptrons finish learning for 192 clock cycles, resulting in a similar accuracy of 98% to the traditional perceptron-based learning algorithm. The proposed learning time estimated using a 65-nm CMOS technology is 0.96 μs, which can achieve around 5,600 x faster learning speed.

The rest of this paper is organized as follows. Section 2 reviews invertible logic and stochastic computing. Section 3 designs the proposed invertible binarized per-

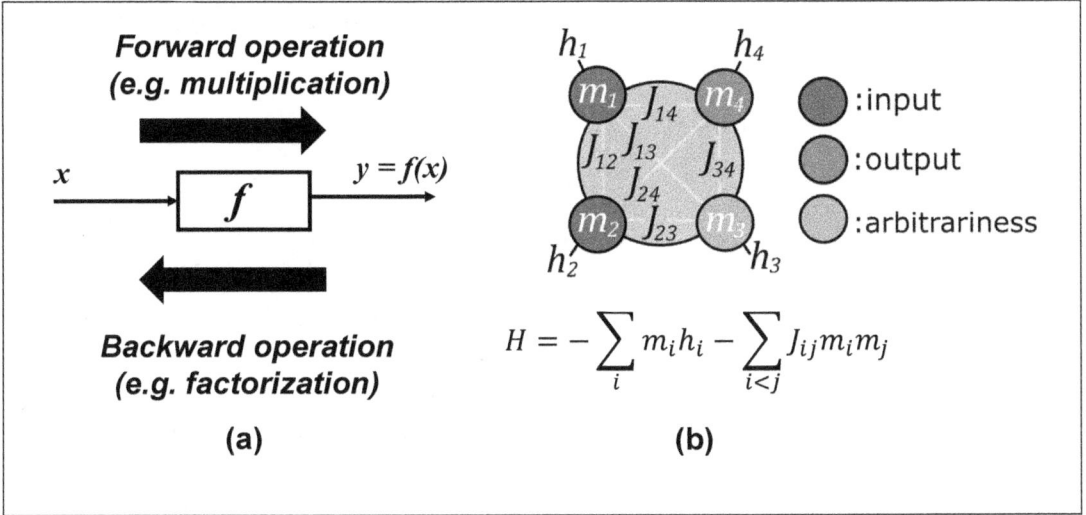

Figure 1: Invertible logic: (a) invertible logic circuit that realizes both forward and backward operations and (b) Hamiltonian of XNOR as an example of a Hamiltonian.

ceptron based on CMOS invertible logic. Section 4 describes the proposed invertible-learning algorithm, and evaluates the accuracy and the learning speed with the traditional perceptron-based learning algorithm. Section 5 concludes this paper.

2 CMOS invertible logic

2.1 Invertible logic

Fig. 1 (a) shows a concept of invertible logic [1] derived from Boltzmann machine [3] Invertible logic circuits operate at forward and/or backward modes, where functions are embedded using a Hamiltonian with inputs ($x_i \in \{0, 1\}$ ($1 \leq i \leq p$)) and outputs ($y_i \in \{0, 1\}$ ($1 \leq i \leq q$)). For example, an invertible multiplier exhibits a capability of multiplication with fixed inputs (forward mode) and factorization with fixed outputs (backward mode). When partial inputs and outputs are fixed, the invertible multiplier operates as division.

Invertible logic is constructed as a network of p-bits (nodes) [1] that are interconnected based on a Hamiltonian. The Hamiltonian represents an energy of the network given by:

$$H = -\sum_i h_i m_i - \sum_{i<j} J_{ij} m_i m_j, \tag{1}$$

where m_i represents an output of a node ($m_i \in \{-1, 1\}$ ($1 \leq i \leq l$)). J_{ij} represents

43

a weight value between nodes and h represents a bias value at a node. Fig. 1 (b) shows a Hamiltonian of an XNOR gate as an example of a Hamiltonian. There are four nodes including two input nodes, an output node, and an arbitrary node. J and h for XNOR are given by:

$$h_{XNOR} = \begin{bmatrix} +1 & +1 & -2 & -1 \end{bmatrix}, \tag{2a}$$

$$J_{XNOR} = \begin{bmatrix} 0 & -1 & +2 & +1 \\ -1 & 0 & +2 & +1 \\ +2 & +2 & 0 & -2 \\ +1 & +1 & -2 & 0 \end{bmatrix}. \tag{2b}$$

The first 2 rows correspond to the input nodes and the third row corresponds to the output node, while the last row correspond to the arbitrary node. Hamiltonians of simple logic gates can be obtained using ground-state spin logic [15, 16].

Given h and J, each node probabilistically calculates the output (m_i) based on the following equations:

$$m_i(t + \tau) = sgn\Big(rnd(-1, +1) + \tanh(I_i(t + \tau))\Big), \tag{3a}$$

$$I_i(t + \tau) = I_0\Big(h_i + \sum_j J_{ij} m_j(t)\Big), \tag{3b}$$

where $rnd(-1, +1)$ is a uniformly distributed random (real) number between -1 and $+1$, sgn is the sign function (with binary $+1$ or -1 outputs) and I_0 is a scaling factor (an inverse *pseudo-temperature*). As m_i is represented in bipolar format, "m_i = +1" and "m_i = -1" correspond to logic values of '1' and '0', respectively. By controlling noise levels using several parameters, such as I_0, H ideally decreases to the global minimum, leading to desired inputs and/or outputs.

As an example of an operation of invertible logic, let us briefly explain how the invertible multiplier operates as division. In this case, the signals provided are one of the inputs and the output. The nodes corresponding to the known signals fix the outputs as +1 or -1 without calculating Eq. (3), where +1/-1 corresponds to 1/0 of binary signal. For example, if four bits (nodes) in 2's complementary format are used to fix a value of '3', the four nodes fix these outputs as (-1,-1,+1,+1). In contrast, the other nodes operate based on Eq. (3) in order to obtain correct results. The detailed operations of invertible logic are summarized in [1].

In [1], the node operations of Eqs. (3a) and (3b) are realized using a probabilistic magneto-resistive model. In addition, they can be approximated using CMOS digital circuits with stochastic computing [7]. This CMOS invertible logic enables us to realize forward/backward operations in FPGAs and ASICs. In this paper, CMOS invertible logic is used to design invertible binarized perceptrons.

44

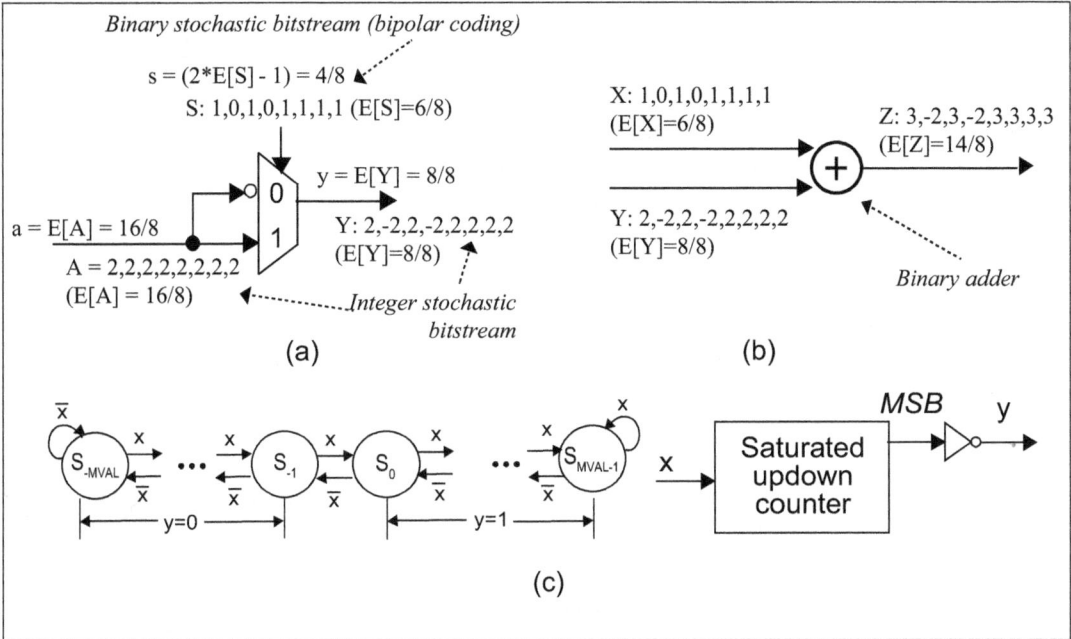

Figure 2: Stochastic circuit blocks: (a) stochastic multiplier of binary and integer stochastic bit streams, (b) stochastic adder, and (c) stochastic tanh (Stan) function.

2.2 Stochastic computing

In stochastic computing, data values are represented by frequencies of '1' in random bit streams [12]. Stochastic computing has been recently applied for area-efficient hardware implementations, such as image processing, error-correcting decoders, deep neural networks [17–24]. It can be categorized into binary stochastic computing and integral stochastic computing [24].

Let us denote by $S \in \{0, 1\}$ a random bit streams in binary stochastic computing. A real number, $s \in [-1 : 1]$, is represented by $(2 * E[S] - 1)$ in binary stochastic computing in bipolar format, where $E[S]$ denotes the expected value of the random variable, S. In case of integral stochastic computing, one or more bit streams are concurrently used to represent data values in larger ranges than that of binary stochastic computing. Let us denote by $X \in \{-r, -(r-1), ..., r\}$ a random bit streams, where $r \in \{1, 2, ...\}$. A real number, $x \in [-r : r]$, is represented by $E[X]$ in signed format, where $E[X]$ denotes the expected value of the random variable X.

Stochastic computing realizes several functions, such as addition, multiplication and nonlinear functions [13]. Fig. 2 (a) shows a stochastic multiplier of an integer stochastic bitstream and a binary stochastic bitstream ($y = a * s$) designed using a two-input multiplexer. In this example, eight bit-length streams are used to represent real values that calculate a multiplication of $16/8 (= a)$ and $4/8 (= s)$, resulting in $8/8 (= y)$. Fig. 2 (b) shows a stochastic adder of both binary/integer stochastic bit streams designed using a binary adder.

Fig. 2 (c) shows a stochastic tanh function block (Stanh) realized using a finite state machine. A tanh function is approximated using Stanh as follows:

$$y = \tanh(x \cdot \text{MVAL}) \approx \text{Stanh}(2 \cdot \text{MVAL}, x), \qquad (4)$$

where $2 \cdot \text{MVAL}$ is the total number of states. The Stanh block is designed using a saturated updown counter corresponding to the finite state machine. The saturated updown counter operates as follows:

$$S(t+1) = \begin{cases} \text{MVAL} - 1 & (S(t) + X(t) \geq \text{MVAL}) \\ -\text{MVAL} & (S(t) + X(t) < -\text{MVAL}) \\ S(t) + X(t) & (otherwise) \end{cases}$$

where $S(t)$ is a state of the updown counter and $X(t)$ is a integral stochastic bit at time, t. Using these circuit blocks, Eqs. (3a) and (3b) are approximated using stochastic computing in CMOS invertible logic.

3 Design of invertible binarized perceptron

3.1 Perceptron

The perceptron [11] is a single-layer neural networks, while a multi-layer perceptron is a class of feed-forwarded neural networks. Recently, deep neural networks have exhibited state-of-the-art results, including image classification and speech recognition [25–27], while the training process requires significantly long time (e.g. several days) [28, 29] because of a complicated learning algorithm [30].

In a conventional perceptron-based learning algorithm [11], weights, $\boldsymbol{w}$, are sequentially updated using input data, $\boldsymbol{x}$, and a true label, t. First, an output, y, is calculated as follows:

$$y = \begin{cases} 1 & (\boldsymbol{x} \cdot \boldsymbol{w} \geq th) \\ -1 & (otherwise) \end{cases} \qquad (5)$$

where th is a threshold value. Using the unipolar format, y can be represented using '0' or '1' instead of '-1' or '+1'. Second, a loss function ($loss$) is calculated as $loss = t - y$. Third, $\boldsymbol{w}$ are updated as follows:

$$\boldsymbol{w_{i+1}} = \boldsymbol{w_i} - \eta \cdot loss \cdot \boldsymbol{x}, \qquad (6)$$

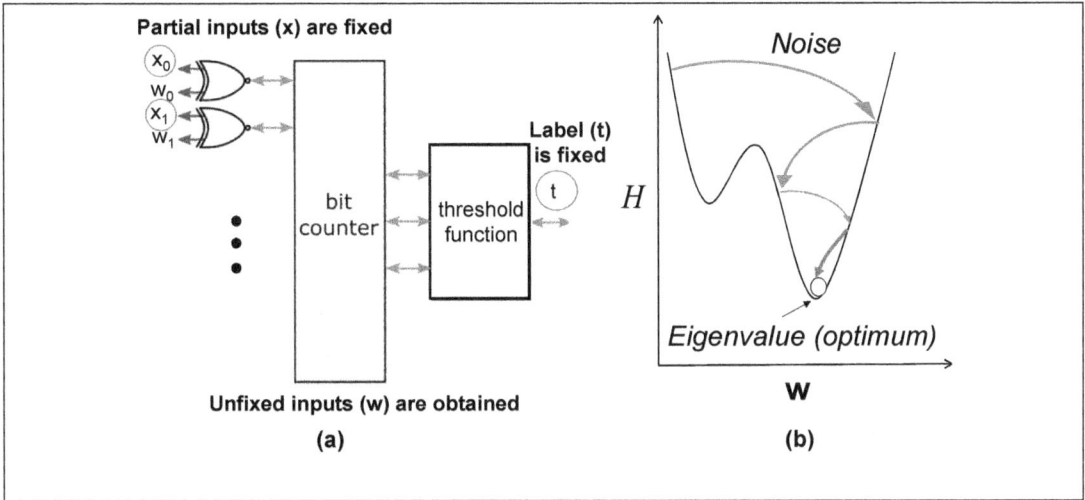

Figure 3: Overview of invertible binarized perceptron: (a) block diagram and (b) reduction of H (energy) by noise. In the invertible binarized perceptron, partial inputs, x, and the true label, t, are fixed to obtain weights, w, based on the energy reduction with noise.

where η is a learning rate and i is the number of iterations. These three processes iteratively operate to update w for increasing the accuracy of perceptron.

3.2 Overview of invertible binarized perceptron

The proposed method learns the perceptron based on CMOS invertible logic without the loss function used in the conventional learning algorithm in Eq. (6). Fig. 3 describes an overview of the invertible perceptron. For the simple hardware implementation, in this paper, x and w are binarized. This binarized perceptron is used as a portion of binary neural networks (BinaryNet) [31], where multiplication used in the conventional perceptrons are replaced by XNOR.

Let us briefly explain the proposed learning flow for the perceptron. When learning w as unknown information, partial inputs, x, and a true label, t are fixed as given information. The unfixed w can be obtained based on the forward/backward operations realized using CMOS invertible logic. The invertible binarized perceptron probabilistically operates using noise signals, reducing the corresponding energy (H) defined by Eq. (1). The H is preliminarily designed so as to set w at the eigenvalue of H, which corresponds to the optimum energy of H. With good noise parameters, H can reach to the optimum energy to obtain w. The detailed learning flow is explained in the following subsections.

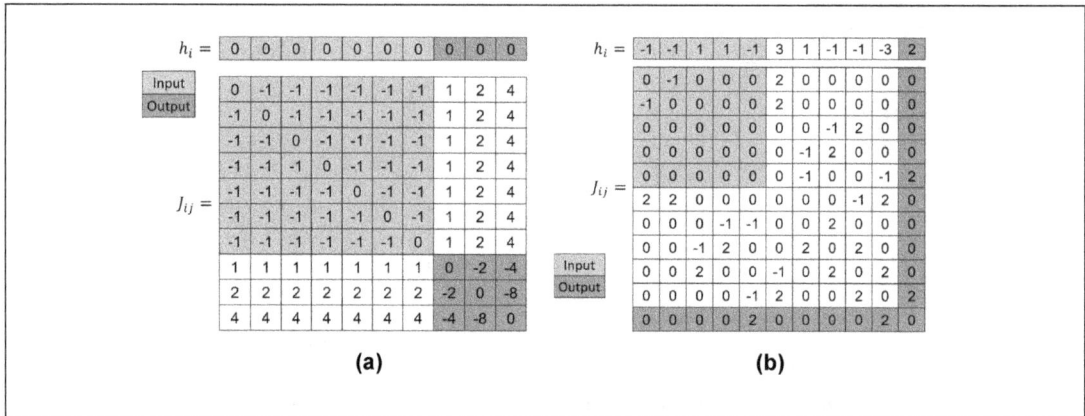

Figure 4: Hamiltonians of: (a) 7-input bit counter and (b) 5-bit threshold function with a threshold value of 13.

3.3 Hamiltonian design

In order to obtain H (Hamiltonian) of the invertible binarized perceptron, Hamiltonians of an XNOR gate, a bit counter, and a threshold function are required as the function of the binarized perceptron is created using these functions as shown in Fig. 3 (a). The Hamiltonian of XNOR is already defined in Eq. (2), where there are two input nodes, an output node, and an arbitrary node. The other Hamiltonians are also obtained using the same method [15, 16] as XNOR.

Fig. 4 (a) shows a Hamiltonian of 7-input bit counter including seven input nodes and three output nodes. The output node is represented by 3 bits, where the last row of the matrix represents the most significant bit (MSB). Fig. 4 (b) shows a Hamiltonian of a 5-bit threshold function with a threshold value of 13, including five input nodes, an output node, and five arbitrary nodes.

The whole Hamiltonian (H_{IBP}) of the binarized perceptron is obtained by superposing these Hamiltonians as follows:

$$H_{IBP} = \sum H_{XNOR} + H_{BC} + H_{TH}, \tag{7}$$

where H_{BC} is the Hamiltonian of the bit counter and H_{TH} is the Hamiltonian of the threshold function. This superposing mechanism is fully described in [7, 15, 16].

3.4 Hardware design using spin-gate circuits

In this paper, we design a 25-input invertible binarized perceptron as shown in Fig. 5 (a). The 25-input bit counter is designed using three 7-input bit counters and

48

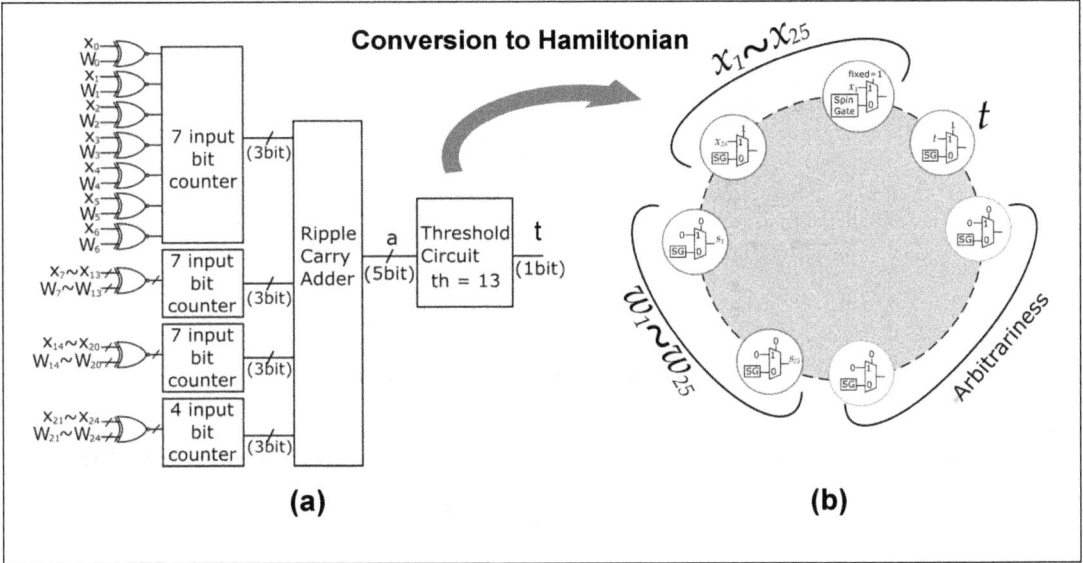

Figure 5: Hardware architecture of the invertible binarized perceptron with 25 inputs: (a) block diagram and (b) corresponding H, where each node is designed using a spin-gate circuit.

a 4-input bit counter with a ripple carry adder. The Hamiltonian (H_{IBP25}) of the 25-input invertible binarized perceptron is created as follows:

$$H_{IBP25} = \sum^{25} H_{XNOR} + \sum^{3} H_{BC7} + H_{BC4} + H_{RCA} + H_{TH}, \qquad (8)$$

where H_{BC7} and H_{BC4} are Hamiltonian of the 7-input and the 4-input bit counter, respectively. H_{RCA} is a Hamiltonian of the ripple carry adder by superposing Hamiltonians of a half adder and a full adder [7].

Fig. 5 (b) shows a description of H_{IBP25}, including 138 nodes in total. There are 25 nodes for $\boldsymbol{x}$, 25 nodes for $\boldsymbol{w}$, a node for t, and 87 arbitrary nodes. The arbitrary nodes are required to connect among input, weight, and output nodes, where the number of arbitrary nodes corresponds to intermediate connections shown in Fig. 5 (a). Each node is designed using a spin-gate circuit as shown in Fig. 6, while each node operates at a fixed mode or an unfixed mode. The spin-gate circuit designed using binary and integral stochastic computing approximates the original equation of Eq. (3) as follows:

$$m_i\left(t + \tau\right) \simeq \mathrm{sgn}\Big(\tanh\!\big(\mathrm{I}_i\left(t + \tau\right) \cdot \mathrm{MVAL}\big)\Big), \qquad (9a)$$

49

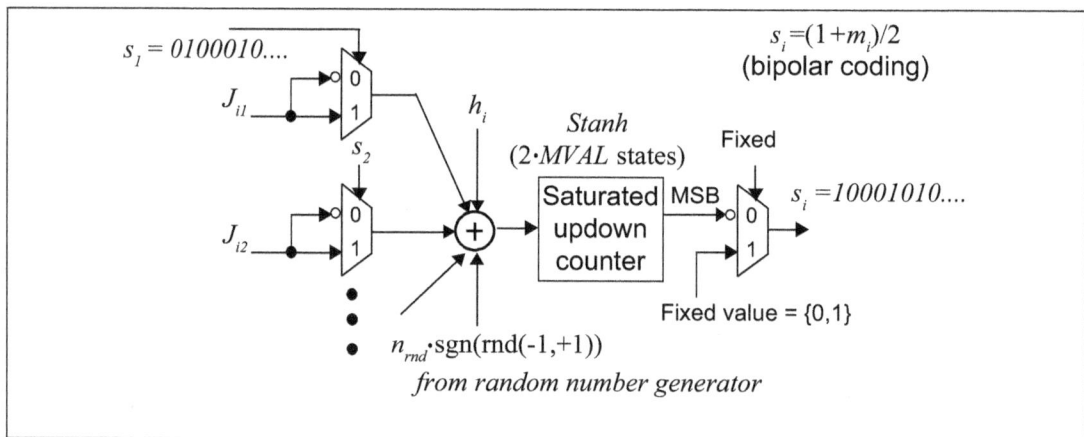

Figure 6: Spin-gate circuit designed using stochastic computing based on Eq. (9).

$$I_i\left(t+\tau\right) \simeq \left(h_i + \sum_j J_{ij}m_j\left(t\right) + n_{rnd} \cdot \text{sgn}\Big(\text{rnd}(-1,+1)\Big)\right), \qquad (9b)$$

where the weighted noise source with a corresponding magnitude denoted as n_{rnd} is an additional parameter from Eq. (3). The weighted noise source is generated using an xor-shift random number generator [32]. The inputs and the output of the spin-gate circuits (m_i) are represented in binary stochastic computing in bipolar format as stochastic bit streams, $s_i = (1 + m_i)/2$. Instead, integral stochastic computing is exploited inside the spin-gate circuits in order to deal with integer values of h and J. As the circuit is fully designed using stochastic computing, it is synthesizable for standard digital CMOS circuits.

When learning the binarized perceptron, $\boldsymbol{w}$ are obtained using $\boldsymbol{x}$ and t. The spin-gate circuits for $\boldsymbol{x}$ and t operate at the fixed mode because these information are given. At the fixed mode, the output of the spin-gate circuit is fixed by the output multiplexer as '0' or '1'. In contrast, the spin-gate circuits for $\boldsymbol{w}$ and the arbitrary nodes operate at the unfixed mode. At the unfixed mode, the spin-gate circuit operates based on Eq. (9). The outputs of the spin-gate circuits for $\boldsymbol{w}$ are random bit streams of $\boldsymbol{s_w}$, which are used for obtaining $\boldsymbol{w}$. The detailed learning algorithm is described in the next section.

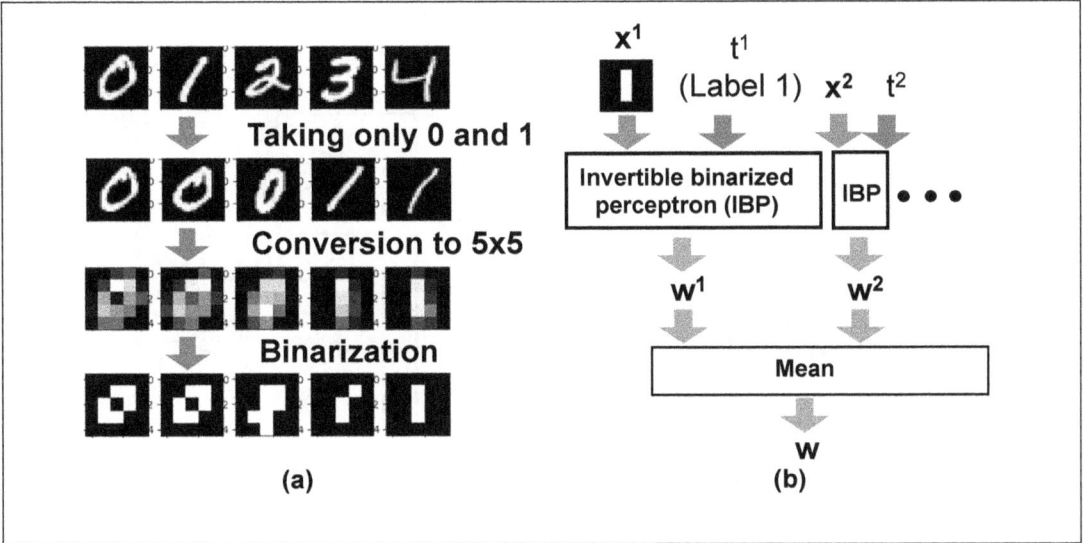

Figure 7: Learning the binarized perceptron using a simplified MNIST dataset: (a) a binarized MNIST with of 5x5 pixel images and (b) the proposed learning process using the parallel invertible binarized perceptrons (IBPs).

4 Evaluation

4.1 Invertible-learning algorithm

The 25-input binarized perceptron is learned using a simplified MNIST dataset as shown in Fig. 7 (a). The simplified MNIST dataset is created as following. First, images and labels of '0' and '1' are extracted from the MNIST database [14]. Second, these images are shrunk to 5x5-pixel images. Third, these grayscale images are converted to binarized images. Fourth, 100 training and 100 test images with labels are randomly selected.

In the proposed learning method, the 100 training data are used in parallel to obtain w using the invertible binarized perceptrons (IBPs) as shown in Fig. 7 (b). The proposed IBP is designed using SystemVerilog and is simulated using Verilator [33]. Each IBP generates w^n $(1 \leq n \leq 100)$ using the training data of x^n and t^n. Then, w is obtained by averaging the 100 temporal weights, w^n, evaluating the accuracy of the binarized perceptron.

At each IBP, w^n is obtained based on the proposed invertible-learning algorithm as shown in Fig. 8. Note that the superscript of n is removed in this paragraph for the simple explanation. Given x and t, a noise signal controlled by n_{rnd} of Eq. (9b) is added. n_{rnd} is parameterized using RND_WEIGHT and T as shown in Fig. 9. Each

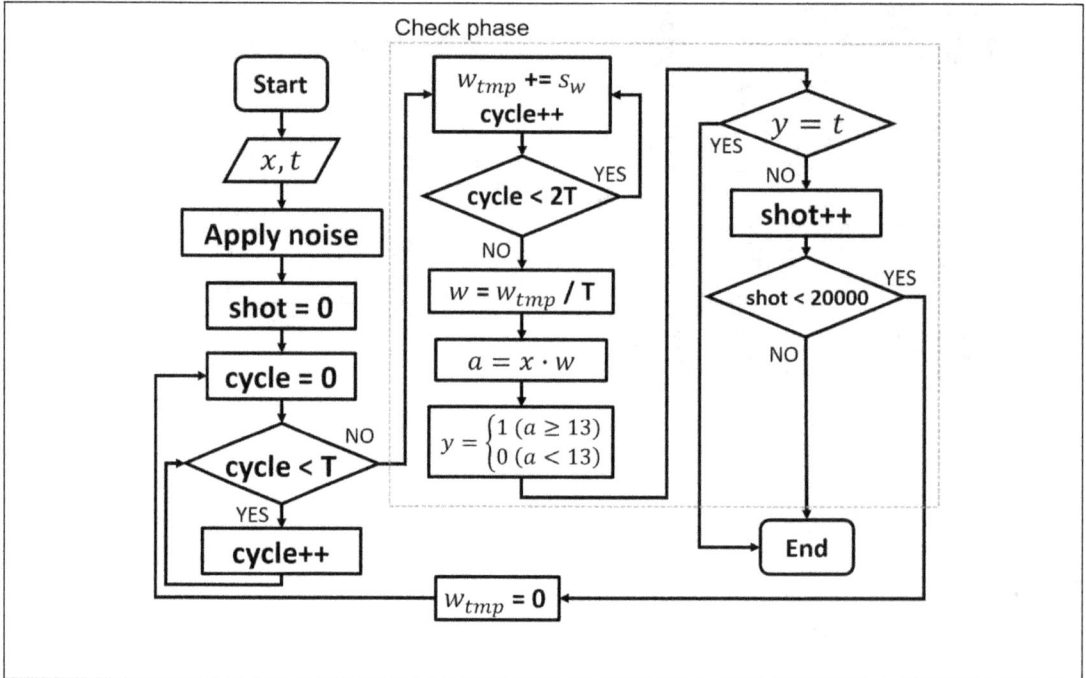

Figure 8: Proposed invertible-learning algorithm at each IBP. The superscript of n is removed for the simple explanation in this figure.

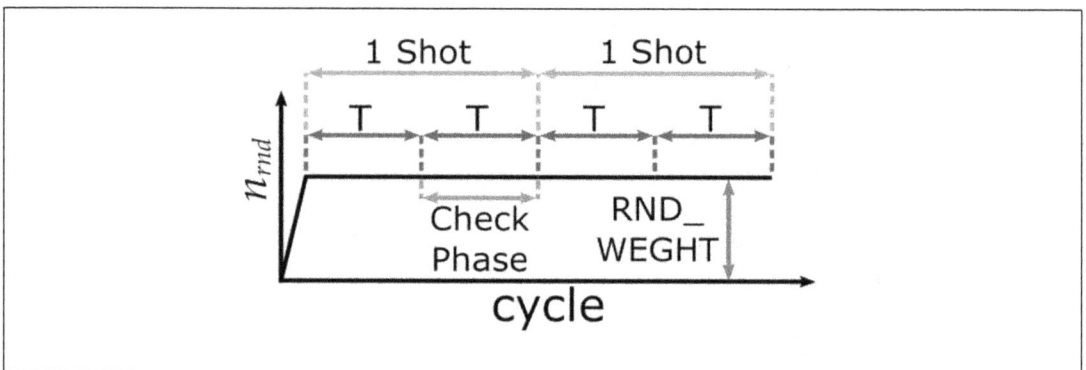

Figure 9: Parameterized magnitude, n_{rnd}, for noise signals used in Fig. 6. Each shot contains $2T$ cycles, where s_w is checked at the last T cycles for obtaining w.

	T	RND_WEIGHT	MVAL	N_{shot}	N_{cycle}	Accuracy [%]
C1	12	17	21	8	192	98
C2	16	23	24	6	192	99
C3	18	15	14	16	576	98
C4	19	15	22	5	190	98
C5	25	21	21	24	1200	98

Table 1: Optimized parameter combination for IBPs with the highest accuracy of 98%.

shot contains the first T cycles and the second T cycles for checking $s_w \in \{0,1\}$, which are the output random bit streams of the spin-gate circuits for $\boldsymbol{w}$. During the last T cycles at each shot, $\boldsymbol{w_{tmp}}$ is accumulated using $\boldsymbol{s_w}$, which then obtains $\boldsymbol{w}$ as $\boldsymbol{w_{tmp}}/T$. After the operation of IBP during $2T$ cycles, a label, y, is calculated using binary logic based on Eq. (5) to check whether t is equal to y. If yes, the learning is complete to determine $\boldsymbol{w}$. Otherwise, the learning process continues until the condition of $t = y$ is satisfied or the number of shots is 20,000.

Each IBP finishes the operation at different number of shots because of different training data and label. N_{shot} is defined as the maximum number of shots among the 100 IBPs. As $\boldsymbol{w}$ is obtained by averaging the 100 $\boldsymbol{w^n}$, the total number of clock cycles, N_{cycle}, is defined as follows

$$N_{cycle} = 2 \cdot T \cdot N_{shot}. \tag{10}$$

In order to minimize N_{cycle} with a high accuracy, we sweep three parameters, T, RND_WEIGHT, and $MVAL$. T and RND_WEIGHT are used to control n_{rnd} as shown in Fig. 9. $MVAL$ is used to control the number of states in the saturated updown counter corresponding to Stanh as shown in Fig. 6. Table 1 summarizes optimized parameter combination for IBPs with the highest accuracy of 98%.

4.2 Performance comparisons

The proposed invertible-learning algorithm is compared with the conventional learning algorithm [11]. The conventional perceptron-based learning algorihtm in Eq. (6) is designed for the 25-input binarized perceptron using Python, where the simulation environment is AMD Ryzen 2990WX (3 GHz) and a 64 GB memory on Ubuntu 18.04.

Fig. 10 illustrates the accuracy vs. iteration (number of training data) in the conventional and the proposed learning algorithms. The conventional algorithm

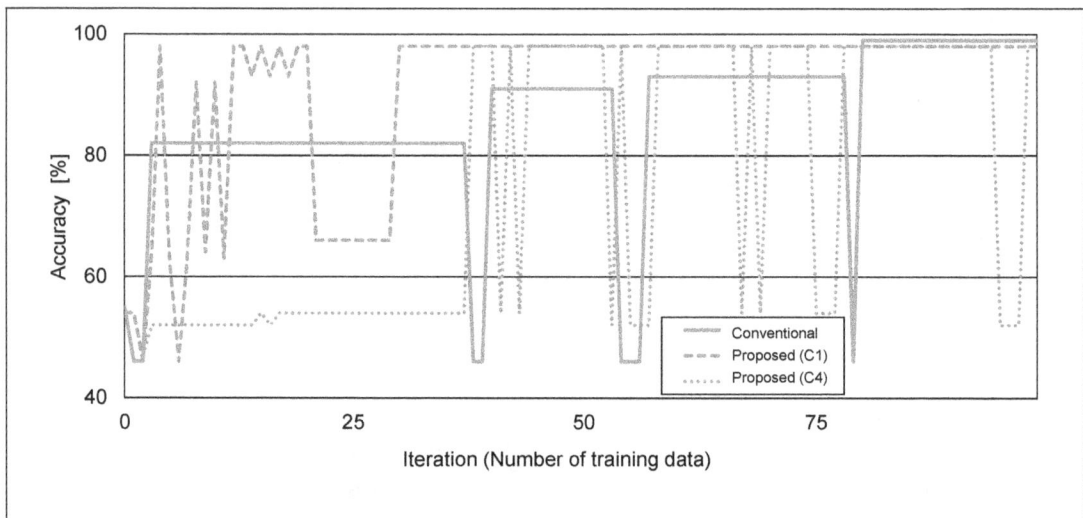

Figure 10: Accuracies of binarized perceptrons using the conventional and the proposed learning algorithms.

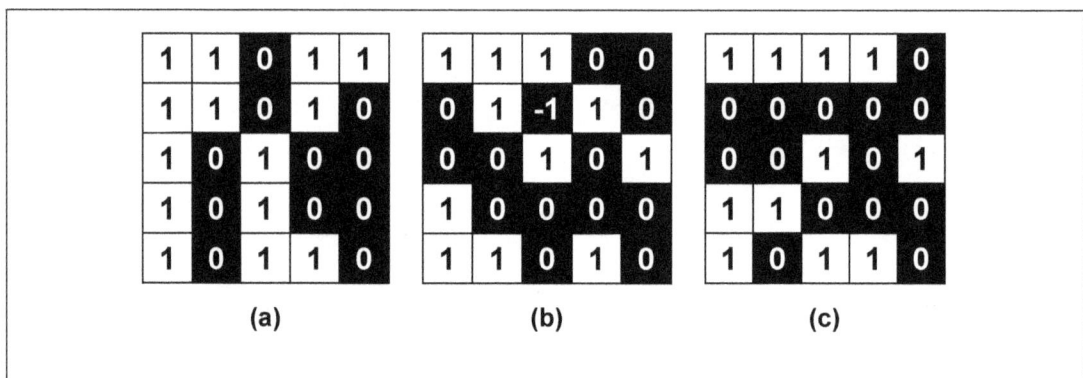

Figure 11: Obtained weights with 100 training data: (a) conventional, (b) proposed (C1), and (c) proposed (C4).

achieves a 99% accuracy with $\eta = 0.5$ after 100 iterations, while the proposed method achieves a 98% accuracy. In the proposed method, C1 and C4 are selected for achieving the lowest and the 2nd lowest N_{cyclc}, respectively. The accuracy of C1 quickly reaches to 98% after training 30 data, while that of C4 is frequently changed during training. The final weights, w, are summarized in Fig. 11.

Table 2 compares the performance of training the 25-input binarized perceptron. The conventional perceptron-based algorithm requires the sequential process

	Conventional [11]	Proposed
Learning algorithm	Perceptron-based learning	Invertible learning
Learning process	Sequential	Parallel
Computing	Floating-point arithmetic	Stochastic computing
Accuracy	99%	98%
Learning time	5.34 ms	0.96 μs
N_{cycle}	-	192

Table 2: Performance comparisons of learning binarized perceptrons using 100 training data and 100 test data of the simplified MNIST dataset.

of Eq. (6) realized based on typical binary logic and floating-point arithmetic in software, resulting in the learning time of 5.34 ms. In contrast, the proposed algorithm can operate in parallel as illustrated in Fig. 7 (b) using stochastic-computing based hardware as shown in Fig. 6. The number of clock cycles, N_{cycle}, is 192 for the proposed (C1), which corresponds to the learning time of 0.96 μs, where the clock frequency of the invertible circuit is 200 MHz using a 65-nm CMOS technology [7]. As a result, the proposed hardware-based learning algorithm can achieve around a 5,600 x faster learning speed than the conventional perceptron-based algorithm, while maintaining a similar accuracy of 98%.

In addition, the conventional algorithm is designed using Verilog HDL with 32-bit fixed-point precisions for performance comparisons in hardware. The hardware-based classical approach takes 5,150 clock cycles with a clock frequency of 100 MHz in FPGA (Digilent Genesys 2), which results in a learning time of 51.2 μs. Compared with the hardware-based conventional approach, the proposed method is about 50-times faster. Note that the conventional algorithm using a loss function requires floating-point computation for training neural networks with high accuracies. Therefore, it is difficult to implement the conventional algorithm for neural networks in hardware, unlike the training for the binary perceptron. The proposed method would be more efficient in learning time for training neural networks.

5   Conclusion

In this paper, we have presented the fast hardware-based learning algorithm for the binarized perceptron using CMOS invertible logic. The proposed invertible-learning algorithm directly obtains the weights without calculating the loss function used in the conventional algorithm, thanks to the bidirectional-computing capability of CMOS invertible logic. This results in the parallel learning process realized using stochastic-computing based hardware as opposed to the conventional sequential learning algorithm. For performance comparison, the 25-input binarized perceptron is learned using the simplified MNIST dataset. The proposed learning speed estimated using a 65-nm CMOS technology can be around a 5,600 x faster than the traditional perceptron-based learning algorithm, while maintaining a similar accuracy of 98%.

Currently, the proposed hardware-based learning algorithm has been designed for the perceptron. In the future prospect, we will apply the learning algorithm for neural networks, which could result in a fast hardware-based learning in comparison with a traditional gradient-descent algorithm.

Acknowledgment

This work was supported in part by JST PRESTO Grant Number JPMJPR18M5 and MEXT Brainware LSI Project.

References

[1] Kerem Camsari, Rafatul Faria, Brian Sutton, and Supriyo Datta. Stochastic p-bits for invertible logic. *Physical Review X*, 7, July 2017.

[2] Kerem Y. Camsari, Brian M. Sutton, and Supriyo Datta. p-bits for probabilistic spin logic. *Applied Physics Reviews*, 6(1):011305, 2019.

[3] G. E. Hinton, T. J. Sejnowski, and D. H. Ackley. Boltzmann machines: Constraint satisfaction networks that learn. Technical Report CMU-CS-84-119, Department of Computer Science, Carnegie-Mellon University, 1984.

[4] R. Faria, K. Y. Camsari, and S. Datta. Low-barrier nanomagnets as p-bits for spin logic. *IEEE Magnetics Letters*, 8:1–5, 2017.

[5] A. De Vos. *Reversible Computing: Fundamentals, Quantum Computing and Applications*. Wiley-VCH Verlag, 2010.

[6] M. Saeedi and I. L. Markov. Synthesis and optimization of reversible circuits—a survey. *ACM Comput. Surv.*, 45(2):21:1–21:34, March 2013.

[7] S. C. Smithson, N. Onizawa, B. H. Meyer, W. J. Gross, and T. Hanyu. Efficient CMOS invertible logic using stochastic computing. *IEEE Transactions on Circuits and Systems I: Regular Papers*, 66(6):2263–2274, June 2019.

[8] J. V. Monaco and M. M. Vindiola. Factoring integers with a brain-inspired computer. *IEEE Transactions on Circuits and Systems I: Regular Papers*, 65(3):1051–1062, March 2018.

[9] Z. Zhou et al. Energy-efficient optimization for concurrent compositions of wsn services. *IEEE Access*, 5:19994–20008, 2017.

[10] A. Zeeshan Pervaiz, B. M. Sutton, L. Anirudh Ghantasala, and K. Y. Camsari. Weighted p-bits for FPGA implementation of probabilistic circuits. *ArXiv e-prints*, Dec. 2017.

[11] S. I. Gallant. Perceptron-based learning algorithms. *IEEE Transactions on Neural Networks*, 1(2):179–191, June 1990.

[12] B. R. Gaines. Stochastic computing. In *Proceedings of the April 18-20, 1967, Spring Joint Computer Conference*, AFIPS '67 (Spring), pages 149–156. ACM, 1967.

[13] B.D. Brown and H.C. Card. Stochastic neural computation I: Computational elements. *IEEE Trans. Comput.*, 50(9):891–905, Sept. 2001.

[14] Y. Lecun and C. Cortes. The MNIST database of handwritten digits. `http://yann.lecun.com/exdb/mnistl/`.

[15] J. D. Biamonte. Non-perturbative k-body to two-body commuting conversion Hamiltonians and embedding problem instances into ising spins. *Physical Review A*, 77:052331, 2008.

[16] J. D. Whitfield, M. Faccin, and J. D. Biamonte. Ground-state spin logic. *Europhysics Letters*, 99(5):57004, 2012.

[17] L. Peng and D. J. Lilja. Using stochastic computing to implement digital image processing algorithms. In *29th ICCD*, pages 154–161, Oct 2011.

[18] A. Alaghi, C. Li, and J. P. Hayes. Stochastic circuits for real-time image-processing applications. In *50th DAC*, pages 1–6, May 2013.

[19] P. Li, D. J. Lilja, W. Qian, K. Bazargan, and M. D. Riedel. Computation on stochastic bit streams digital image processing case studies. *IEEE Transactions on Very Large Scale Integration (VLSI) Systems*, 22(3):449–462, Mar. 2014.

[20] P. Li, D. J. Lilja, W. Qian, M. D. Riedel, and K. Bazargan. Logical computation on stochastic bit streams with linear finite-state machines. *IEEE Trans. Comput.*, 63(6):1474–1486, June 2014.

[21] S. S. Tehrani, S. Mannor, and W. J. Gross. Fully parallel stochastic LDPC decoders. *IEEE Trans. on Signal Processing*, 56(11):5692–5703, Nov. 2008.

[22] S. S. Tehrani, A. Naderi, G. A. Kamendje, S. Hemati, S. Mannor, and W. J. Gross. Majority-based tracking forecast memories for stochastic LDPC decoding. *IEEE Trans. on Signal Processing*, 58(9):4883–4896, Sep. 2010.

[23] J. Chen, J. Hu, and J. Zhou. Hardware and energy-efficient stochastic LU decomposition scheme for MIMO receivers. *IEEE Trans. on Very Large Scale Integration (VLSI) Systems*, 24(4):1391–1401, April 2016.

[24] A. Ardakani, F. Leduc-Primeau, N. Onizawa, T. Hanyu, and W. J. Gross. VLSI implementation of deep neural network using integral stochastic computing. *IEEE Trans. on Very Large Scale Integration (VLSI) Systems*, 2017 (to apeear).

[25] P. Luo, Y. Tian, X. Wang, and X. Tang. Switchable deep network for pedestrian detection. In *2014 IEEE Conference on Computer Vision and Pattern Recognition*, pages 899–906, June 2014.

[26] X. Zeng, W. Ouyang, and X. Wang. Multi-stage contextual deep learning for pedestrian

detection. In *2013 IEEE International Conference on Computer Vision*, pages 121–128, Dec 2013.

[27] L. Chen, G. Papandreou, I. Kokkinos, K. Murphy, and A. L. Yuille. Deeplab: Semantic image segmentation with deep convolutional nets, atrous convolution, and fully connected crfs. *IEEE Transactions on Pattern Analysis and Machine Intelligence*, 40(4):834–848, April 2018.

[28] M. Abadi et al. Tensorflow: Large-scale machine learning on heterogeneous distributed systems. *ArXiv e-prints*, Mar. 2016.

[29] D. Li, X. Chen, M. Becchi, and Z. Zong. Evaluating the energy efficiency of deep convolutional neural networks on cpus and gpus. In *2016 IEEE International Conferences on Big Data and Cloud Computing (BDCloud), Social Computing and Networking (SocialCom), Sustainable Computing and Communications (SustainCom) (BDCloud-SocialCom-SustainCom)*, pages 477–484, Oct 2016.

[30] J. Duchi, E. Hazan, and Y. Singer. Adaptive subgradient methods for online learning and stochastic optimization. *Journal of Machine Learning Research*, 12:2121–2159, 2011.

[31] M. Courbariaux and Y. Bengio. BinaryNet: Training deep neural networks with weights and activations constrained to +1 or -1. *CoRR*, abs/1602.02830, 2016.

[32] Sebastiano Vigna. Further scramblings of Marsaglia's xorshift generators. *Journal of Computational and Applied Mathematics*, 315(Supplement C):175–181, 2017.

[33] W. Snyder, P. Wasson, and D. Galbi. Verilator: Convert verilog code to c++/systemc. `https://www.veripool.org/wiki/verilator`, 2012.

 Received 30 April 2019

SOME PROPERTIES OF FIRST-ORDER NELSONIAN PARACONSISTENT QUANTUM LOGIC

NORIHIRO KAMIDE
Teikyo University,
Faculty of Science and Engineering,
Department of Information and Electronic Engineering,
Toyosatodai 1-1, Utsunomiya, Tochigi 320-8551, Japan
`drnkamide08@kpd.biglobe.ne.jp`

Abstract

In this study, several desirable properties of a novel logic called first-order Nelsonian paraconsistent quantum logic are investigated. To investigate such properties, a single-antecedent/succedent sequent calculus NL for this logic is introduced. This logic is regarded as a combination of both Nelson's paraconsistent four-valued logic and Dalla Chiara and Giuntini's paraconsistent quantum logic. The duality and cut-elimination theorems for NL are proved. The decidability, several constructive properties, several constructible falsity properties, the Craig interpolation property, and several uniform and dual-uniform provabilities for NL are shown. The constructive properties that are studied consist of the conjunction and universal properties, which are the duals of the standard disjunction and existence properties, respectively. The constructible falsity properties that are discussed consist of the constructible falsity properties with respect to negated disjunction and negated existential quantifier. These constructive and constructible falsity properties and the dual-uniform provability are regarded as characteristic properties of NL, because these properties do not hold for the standard paraconsistent logics. Furthermore, an extension of NL is constructed from NL by adding several naive comprehension rules from the naive set theory. It is shown that this extended system is consistent, and Russell's (like) paradox does not imply the inconsistency of this system.

This paper is an extension of the paper presented in the 49th IEEE International Symposium on Multiple-Valued Logic [21].

1 Introduction

In this study, several desirable properties of a novel logic called *first-order Nelsonian paraconsistent quantum logic* are investigated. This logic is a combination of *Nelson's paraconsistent four-valued logic* [1, 28, 22] and *Dalla Chiara and Giuntini's paraconsistent quantum logic* [10]. Nelson's paraconsistent four-valued logic is a natural extension of *Belnap and Dunn's paraconsistent four-valued logic* (or *first-degree entailment logic*) [12, 2, 3]. Dalla Chiara and Giuntini's paraconsistent quantum logic is a common denominator of the *minimal quantum logic* (or *orthologic*) [5] and Belnap and Dunn's paraconsistent four-valued logic. While *paraconsistent logics* [29] are useful for representing inconsistency-tolerant reasoning, *quantum logics* [5] are suitable for describing quantum phenomena. The proposed first-order Nelsonian paraconsistent quantum logic is therefore suitable for representing and describing both inconsistency-tolerant reasoning and quantum phenomena.

The original quantum logic, which was described as a logical abstraction from the algebraic structure of the closed subspaces in a Hilbert space, was introduced by Birkhoff and von Neumann [5]. The paraconsistent quantum logic, a weak form of quantum logic for which the non-contradiction and excluded-middle laws do not hold, was introduced by Dalla Chiara and Giuntini [10]. These authors developed an axiomatization, a Kripke-type semantics, and an algebraic semantics for two forms of this logic. A cut-free sequent calculus for this logic was introduced by Faggian and Sambin [13], and these authors also referred to the paraconsistent quantum logic as *basic orthologic* in [13]. Several alternative cut-free sequent calculi for the paraconsistent quantum logic were also introduced by Kamide in [19]. However, several properties (such as Craig interpolation and uniform probability) of some first-order and/or modified extensions of this logic have not yet been investigated.

To investigate several desirable properties (and theorems) of the proposed first-order Nelsonian paraconsistent quantum logic, we introduce a simple *single-antecedent/succedent sequent calculus* NL for it. This calculus NL is an extension of the single-antecedent/succedent sequent calculus PQL proposed in [19] for the paraconsistent quantum logic. The single-antecedent/succedent sequents $\Gamma \Rightarrow \Delta$ of NL have the restriction that both Γ and Δ consist of either a single formula or the empty set rather than a set of formulas, i.e., $\Gamma \Rightarrow \Delta$ is of the form $\gamma \Rightarrow \delta$, where γ and δ are either a single formula or the empty set. This restriction on these sequents is regarded as the intersection of the sequent restrictions on the standard sequent calculi for *intuitionistic logic* and *dual-intuitionistic logic* [9, 16, 30]. The sequent calculus introduced by Faggian and Sambin for the paraconsistent quantum logic is based on the standard sequent $\Gamma \Rightarrow \Delta$, where Γ and Δ consist of a set of formulas. It was thus shown in [19] that the standard non-restricted sequents $\Gamma \Rightarrow \Delta$ of paraconsistent quantum logic can be restricted to the single antecedent/succedent sequents $\gamma \Rightarrow \delta$.

Several single-antecedent/succedent sequent calculi for logics without Nelsonian paraconsistent negation have been studied. A comprehensive survey of such systems appear in [19]. More recently, a single-antecedent/succedent sequent calculus, $S11_{\wedge,\vee}$, which corresponds to the conjunction-disjunction fragment of NL, has been studied by Beziau [4] from the point of view of fibring logics and lattice theory. In [4], the author referred to this single-antecedent/succedent sequent calculus as a *monosequent proof system*.

Use of the single antecedent/succedent sequents of NL considerably simplifies the proof of the cut-elimination and decidability of NL. In particular, the double induction on grade and rank is not needed in the cut-elimination proof. By using the restricted sequents, several characteristic properties such as the duality, conjunction, and universal properties are easily obtained. These results for obtaining the characteristic properties are a significant contribution of this paper. Moreover, as mentioned in a later section of this paper, the sequent restriction is also useful for proving the cut-elimination theorem for an extended NL with some naive comprehension rules from the naive set theory.

A propositional fragment NQL of NL was introduced and studied in [20], and the cut-elimination, decidability and duality theorems for NQL were shown. In [20], several first-order, modal, temporal, and infinitary extensions of NQL were also considered, and several theorems for embedding these extensions into their negation-free fragments and a theorem for embedding the temporal extension into the infinitary extension were proved.

However, a weak negation connective $\neg$, which can be defined by the implication connective $\rightarrow$ with the falsity constant $\bot$, and a weak co-negation connective $\neg'$, which can be defined by the co-implication connective $\leftarrow$ with the truth constant $\top$, were not considered in [20]. For more information on constructive negation, constructive co-negation, constructive implication, and constructive co-implication, see e.g., [31]. Several constructive and constructible falsity properties for these systems, which are concerned with $\neg$ and $\neg'$ as well as a weak paraconsistent negation connective $\sim$, were not investigated in [20]. In addition, the author of [20] did not prove the Craig interpolation and uniform and dual-uniform provability theorems for NQL, although the uniform provability theorem is known to be important for obtaining a foundation for sequent calculus-based logic programming languages [27]. Thus, the present paper considers an extended system NL that has three types of negation connectives $\neg$, $\neg'$, and $\sim$, and shows the above-mentioned characteristic and useful properties and theorems of the proposed first-order Nelsonian paraconsistent logic.

The contents of this paper are as follows.

In Section 2, the sequent calculus NL, which is an extension of PQL and NQL, is introduced, and the duality theorem for NL is proved. The duality theorem is a characteristic theorem of NL, because this theorem does not hold for standard paraconsistent logics. For more information on paraconsistent logics, see e.g., [29] and the references therein.

In Section 3, the cut-elimination theorem for NL is proved, and some applications of this theorem are presented. These applications include the decidability, the paraconsistency, several constructive properties, several constructible falsity properties, and the Craig interpolation property. The constructive properties that are discussed include the conjunction and universal properties, which are the duals of the standard disjunction and existence properties, respectively. The constructible falsity properties that are discussed include the constructible falsity properties with respect to negated disjunction and negated existential quantifier, which are the duals of the original constructible falsity properties of Nelson [28]. It is noted that these constructive and constructible falsity properties are also regarded as characteristic properties of NL, because they do not hold for standard paraconsistent logics.

In Section 4, the uniform and dual-uniform provability theorems for NL are proved using the permutability theorem for NL. The dual-uniform provability theorem is also regarded as a characteristic theorem for NL, because this theorem does not hold for typical non-classical logics. The permutability theorem for Gentzen's LK [14] for classical logic was originally proved by Kleene [23], the uniform provability theorem for Gentzen's LJ [14] for intuitionistic logic was originally proved by Miller et al. [27], and the uniform provability theorem for a sequent calculus for Nelson's paraconsistent four-valued logic was proved by Kamide in [18].

In Section 5, an extension $NL_\in$ of NL by adding several naive comprehension rules from the naive set theory is investigated. It is shown in this section that the extended system $NL_\in$ is consistent, and Russell's (like) paradox does not imply the inconsistency of $NL_\in$. This fact is similar to that for *Grishin's contraction-less logic* [17], because NL has no contraction rule.

In Section 6, this paper is concluded with some remarks. We present some backgrounds of the original quantum logic by Birkhoff and von Neumann [5] and the paraconsistent quantum logic by Dalla Chiara and Giuntini [10]. Finally, we also illustrate some closely related works on the paraconsistent quantum logic and some related logics.

2 Logic and its duality

Formulas of *first-order Nelsonian paraconsistent quantum logic* are obtained from free variables, bound variables, functions, predicates, logical constants: $\top$ (truth constant) and $\bot$ (falsity constant), logical connectives: $\wedge$ (conjunction), $\vee$ (disjunction), $\rightarrow$ (implication), $\leftarrow$ (co-implication), $\forall$ (any), $\exists$ (exists), $\neg$ (weak negation), $\neg'$ (weak co-negation) and $\sim$ (weak Nelsonian negation). The numbers of free and bound variables are assumed to be countable, and the numbers of functions and predicates are also assumed to be countable. It is also assumed that there is at least one predicate. A 0-ary function is an individual constant, and a 0-ary predicate is a propositional variable. Small letters $p, q, r, \ldots$ are used to denote atomic formulas or predicate symbols, Greek lower-case letters $\alpha, \beta, \ldots$ are used to denote formulas, and small letters $x, y, \ldots$ are used to denote individual variables. An expression $\alpha[y/x]$ means the formula which is obtained from the formula α by replacing all free occurrences of the individual variable x in α with the individual variable y, but avoiding a clash of variables by a suitable renaming of bound variables. A *single-antecedent/succedent sequent* (or simply called *sequent*) is an expression of the form $\gamma \Rightarrow \delta$ where γ and δ are a formula or the empty set, i.e., $\alpha \Rightarrow \beta$, $\alpha \Rightarrow$, $\Rightarrow \beta$ and $\Rightarrow$ are sequents. An expression $\alpha \Leftrightarrow \beta$ is used to represent the sequents $\alpha \Rightarrow \beta$ and $\beta \Rightarrow \alpha$. An expression $L \vdash S$ means that a sequent S is provable in a sequent calculus L. We sometimes omit L in this expression if L is clear from the context. A rule R of inference is said to be *admissible* in a sequent calculus L if the following condition is satisfied: For any instance

$$\frac{S_1 \cdots S_n}{S}$$

of R, if $L \vdash S_i$ for all i, then $L \vdash S$.

A sequent calculus NL for the first-order Nelsonian paraconsistent quantum logic is defined below.

Definition 2.1 (NL). *In the following, γ and δ in the sequents appearing in the initial sequents and inference rules represent a formula or the empty set.*

The initial sequents of NL are of the following form, for any atomic formula p:

$$\gamma \Rightarrow \top \qquad \bot \Rightarrow \delta \qquad \sim\top \Rightarrow \delta \qquad \gamma \Rightarrow \sim\bot \qquad p \Rightarrow p \qquad \sim p \Rightarrow \sim p.$$

The structural inference rules of NL are of the form:

$$\frac{\gamma \Rightarrow \alpha \quad \alpha \Rightarrow \delta}{\gamma \Rightarrow \delta} \text{ (cut)} \qquad \frac{\Rightarrow \delta}{\alpha \Rightarrow \delta} \text{ (we-left)} \qquad \frac{\gamma \Rightarrow}{\gamma \Rightarrow \alpha} \text{ (we-right).}$$

The non-~-prefixed logical inference rules of NL *are of the form:*

$$\frac{\alpha \Rightarrow \delta}{\alpha \wedge \beta \Rightarrow \delta} \ (\wedge\text{left1}) \quad \frac{\beta \Rightarrow \delta}{\alpha \wedge \beta \Rightarrow \delta} \ (\wedge\text{left2})$$

$$\frac{\gamma \Rightarrow \alpha \quad \gamma \Rightarrow \beta}{\gamma \Rightarrow \alpha \wedge \beta} \ (\wedge\text{right}) \quad \frac{\alpha \Rightarrow \delta \quad \beta \Rightarrow \delta}{\alpha \vee \beta \Rightarrow \delta} \ (\vee\text{left})$$

$$\frac{\gamma \Rightarrow \alpha}{\gamma \Rightarrow \alpha \vee \beta} \ (\vee\text{right1}) \quad \frac{\gamma \Rightarrow \beta}{\gamma \Rightarrow \alpha \vee \beta} \ (\vee\text{right2})$$

$$\frac{\Rightarrow \alpha \quad \beta \Rightarrow \delta}{\alpha \to \beta \Rightarrow \delta} \ (\to\text{left}) \quad \frac{\alpha \Rightarrow \beta}{\Rightarrow \alpha \to \beta} \ (\to\text{right})$$

$$\frac{\alpha \Rightarrow \beta \quad \pi \Rightarrow \sigma}{\beta \to \pi \Rightarrow \alpha \to \sigma} \ (\to\text{order})$$

$$\frac{\alpha \Rightarrow \beta}{\alpha \leftarrow \beta \Rightarrow} \ (\leftarrow\text{left}) \quad \frac{\gamma \Rightarrow \alpha \quad \beta \Rightarrow}{\gamma \Rightarrow \alpha \leftarrow \beta} \ (\leftarrow\text{right})$$

$$\frac{\sigma \Rightarrow \pi \quad \beta \Rightarrow \alpha}{\sigma \leftarrow \alpha \Rightarrow \pi \leftarrow \beta} \ (\leftarrow\text{order})$$

$$\frac{\Rightarrow \alpha}{\neg \alpha \Rightarrow} \ (\neg\text{left}) \quad \frac{\alpha \Rightarrow}{\Rightarrow \neg \alpha} \ (\neg\text{right})$$

$$\frac{\beta \Rightarrow \alpha}{\neg \alpha \Rightarrow \neg \beta} \ (\neg\text{contraposition})$$

$$\frac{\Rightarrow \alpha}{\neg' \alpha \Rightarrow} \ (\neg'\text{left}) \quad \frac{\alpha \Rightarrow}{\Rightarrow \neg' \alpha} \ (\neg'\text{right})$$

$$\frac{\beta \Rightarrow \alpha}{\neg' \alpha \Rightarrow \neg' \beta} \ (\neg'\text{contraposition})$$

$$\frac{\alpha[t/x] \Rightarrow \delta}{\forall x \alpha \Rightarrow \delta} \ (\forall\text{left}) \quad \frac{\gamma \Rightarrow \alpha[a/x]}{\gamma \Rightarrow \forall x \alpha} \ (\forall\text{right})$$

$$\frac{\alpha[a/x] \Rightarrow \delta}{\exists x \alpha \Rightarrow \delta} \ (\exists\text{left}) \quad \frac{\gamma \Rightarrow \alpha[t/x]}{\gamma \Rightarrow \exists x \alpha} \ (\exists\text{right})$$

where a is a free variable which must not occur in the lower sequents of ($\forall$right) *and* ($\exists$left), *and t is an arbitrary term.*

The ~-prefixed logical inference rules of NL *are of the form:*

$$\frac{\alpha \Rightarrow \delta}{\sim\sim\alpha \Rightarrow \delta} \ (\sim\sim\text{left}) \quad \frac{\gamma \Rightarrow \alpha}{\gamma \Rightarrow \sim\sim\alpha} \ (\sim\sim\text{right})$$

$$\frac{\sim\alpha \Rightarrow \delta \quad \sim\beta \Rightarrow \delta}{\sim(\alpha \wedge \beta) \Rightarrow \delta} \ (\sim\wedge\text{left}) \quad \frac{\gamma \Rightarrow \sim\alpha}{\gamma \Rightarrow \sim(\alpha \wedge \beta)} \ (\sim\wedge\text{right1})$$

$$\frac{\gamma \Rightarrow \sim\beta}{\gamma \Rightarrow \sim(\alpha \wedge \beta)} \ (\sim\wedge\text{right2})$$

$$\frac{\sim\alpha \Rightarrow \delta}{\sim(\alpha \vee \beta) \Rightarrow \delta} \ (\sim\vee\text{left1}) \qquad \frac{\sim\beta \Rightarrow \delta}{\sim(\alpha \vee \beta) \Rightarrow \delta} \ (\sim\vee\text{left2})$$

$$\frac{\gamma \Rightarrow \sim\alpha \quad \gamma \Rightarrow \sim\beta}{\gamma \Rightarrow \sim(\alpha \vee \beta)} \ (\sim\vee\text{right})$$

$$\frac{\alpha \Rightarrow \delta}{\sim(\alpha\rightarrow\beta) \Rightarrow \delta} \ (\sim\rightarrow\text{left1}) \qquad \frac{\sim\beta \Rightarrow \delta}{\sim(\alpha\rightarrow\beta) \Rightarrow \delta} \ (\sim\rightarrow\text{left2})$$

$$\frac{\gamma \Rightarrow \alpha \quad \gamma \Rightarrow \sim\beta}{\gamma \Rightarrow \sim(\alpha\rightarrow\beta)} \ (\sim\rightarrow\text{right}) \qquad \frac{\sim\alpha \Rightarrow \delta \quad \beta \Rightarrow \delta}{\sim(\alpha\leftarrow\beta) \Rightarrow \delta} \ (\sim\leftarrow\text{left})$$

$$\frac{\gamma \Rightarrow \sim\alpha}{\gamma \Rightarrow \sim(\alpha\leftarrow\beta)} \ (\sim\leftarrow\text{right1}) \qquad \frac{\gamma \Rightarrow \beta}{\gamma \Rightarrow \sim(\alpha\leftarrow\beta)} \ (\sim\leftarrow\text{right2})$$

$$\frac{\alpha \Rightarrow \delta}{\sim\neg\alpha \Rightarrow \delta} \ (\sim\neg\text{left}) \qquad \frac{\gamma \Rightarrow \alpha}{\gamma \Rightarrow \sim\neg\alpha} \ (\sim\neg\text{right})$$

$$\frac{\alpha \Rightarrow \delta}{\sim\neg'\alpha \Rightarrow \delta} \ (\sim\neg'\text{left}) \qquad \frac{\gamma \Rightarrow \alpha}{\gamma \Rightarrow \sim\neg'\alpha} \ (\sim\neg'\text{right})$$

$$\frac{\sim\alpha[a/x] \Rightarrow \delta}{\sim\forall x\alpha \Rightarrow \delta} \ (\sim\forall\text{left}) \qquad \frac{\gamma \Rightarrow \sim\alpha[t/x]}{\gamma \Rightarrow \sim\forall x\alpha} \ (\sim\forall\text{right})$$

$$\frac{\sim\alpha[t/x] \Rightarrow \delta}{\sim\exists x\alpha \Rightarrow \delta} \ (\sim\exists\text{left}) \qquad \frac{\gamma \Rightarrow \sim\alpha[a/x]}{\gamma \Rightarrow \sim\exists x\alpha} \ (\sim\exists\text{right})$$

where a is a free variable which must not occur in the lower sequents of ($\sim\forall$left) and ($\sim\exists$right), and t is an arbitrary term.

Remark 2.2. *We make the following remarks.*

1. *The $\{\top, \bot, \rightarrow, \leftarrow, \neg, \neg'\}$-free fragment of NL is just the single-antecedent/succedent sequent calculus FPQL presented in [19] for the first-order paraconsistent quantum logic.*

2. *The sequents of the form $\alpha \Rightarrow \alpha$ for any formula α are provable in cut-free NL. This fact can be shown by induction on α. To show this, the rules ($\rightarrow$order), ($\leftarrow$order), ($\neg$contraposition) and ($\neg'$contrapositon) are needed.*

3. *The following sequents are provable in cut-free NL: For any formulas α and β,*

   (a) $\sim\top \Leftrightarrow \bot$,

(b) $\sim\!\perp \Leftrightarrow \top$,

(c) $\sim\!(\alpha \wedge \beta) \Leftrightarrow \sim\!\alpha \vee \sim\!\beta$,

(d) $\sim\!(\alpha \vee \beta) \Leftrightarrow \sim\!\alpha \wedge \sim\!\beta$,

(e) $\sim\!(\alpha \rightarrow \beta) \Leftrightarrow \alpha \wedge \sim\!\beta$,

(f) $\sim\!(\alpha \leftarrow \beta) \Leftrightarrow \sim\!\alpha \vee \beta$,

(g) $\sim\!\sim\!\alpha \Leftrightarrow \alpha$,

(h) $\sim\!\neg\alpha \Leftrightarrow \alpha$,

(i) $\sim\!\neg'\alpha \Leftrightarrow \alpha$,

(j) $\sim\!\forall x \alpha \Leftrightarrow \exists x \sim\!\alpha$,

(k) $\sim\!\exists x \alpha \Leftrightarrow \forall x \sim\!\alpha$.

4. *The formulas which just correspond to the sequents $\sim\!(\alpha \rightarrow \beta) \Leftrightarrow \alpha \wedge \sim\!\beta$ are known to be the characteristic axiom schemes of Nelson's paraconsistent four-valued logic [1, 28].*

5. *The connectives $\neg$ and $\neg'$ can be defined as $\neg\alpha := \alpha \rightarrow \perp$ and $\neg'\alpha := \top \leftarrow \alpha$, respectively.*

6. *The following sequents are not provable in cut-free NL: For any atomic formulas p and q,*

(a) $\neg\neg p \Leftrightarrow p$,

(b) $\neg'\neg'p \Leftrightarrow p$,

(c) $\neg\sim\!p \Leftrightarrow p$,

(d) $\neg'\sim\!p \Leftrightarrow p$,

(e) $\neg\neg'p \Leftrightarrow p$,

(f) $\neg'\neg p \Leftrightarrow p$,

(g) $\neg(p \wedge q) \Leftrightarrow \neg p \vee \neg q$,

(h) $\neg(p \vee q) \Leftrightarrow \neg p \wedge \neg q$,

(i) $\neg'(p \wedge q) \Leftrightarrow \neg'p \vee \neg'q$,

(j) $\neg'(p \vee q) \Leftrightarrow \neg'p \wedge \neg'q$.

This fact is guaranteed by the cut-elimination theorem for NL.

Next, we introduce a self-translation for NL, and using this translation, we show the duality theorem for NL.

Definition 2.3. *We fix a common set Φ of atomic formulas. The language $\mathcal{L}$ of NL is defined using Φ, $\top$, $\bot$, $\wedge$, $\vee$, $\rightarrow$, $\leftarrow$, $\neg$, $\neg'$, $\sim$, $\forall$ and $\exists$.*

A mapping f from $\mathcal{L}$ to $\mathcal{L}$ is defined inductively by:

1. *$f(p) := p$ for any $p \in \Phi$,*

2. *$f(\top) := \bot$,*

3. *$f(\bot) := \top$,*

4. *$f(\alpha \wedge \beta) := f(\alpha) \vee f(\beta)$,*

5. *$f(\alpha \vee \beta) := f(\alpha) \wedge f(\beta)$,*

6. *$f(\alpha \rightarrow \beta) := f(\beta) \leftarrow f(\alpha)$,*

7. *$f(\alpha \leftarrow \beta) := f(\beta) \rightarrow f(\alpha)$,*

8. *$f(\neg \alpha) := \neg' f(\alpha)$,*

9. *$f(\neg' \alpha) := \neg f(\alpha)$,*

10. *$f(\sim \alpha) := \sim f(\alpha)$,*

11. *$f(\forall x \alpha) := \exists x f(\alpha)$,*

12. *$f(\exists x \alpha) := \forall x f(\alpha)$.*

We then obtain the following theorem.

Theorem 2.4 (Duality for NL)**.** *Let f be the mapping defined in Definition 2.3. Then we have:*

$$\text{NL} \vdash \gamma \Rightarrow \delta \ \textit{iff} \ \text{NL} \vdash f(\delta) \Rightarrow f(\gamma).$$

Proof. We show only the direction ($\Longrightarrow$) by induction on the proofs P of $\gamma \Rightarrow \delta$ in NL. We distinguish the cases according to the last inference of P, and show some cases.

1. Case ($\rightarrow$order): The last inference of P is of the form:

$$\frac{\alpha \Rightarrow \beta \quad \pi \Rightarrow \sigma}{\beta \rightarrow \pi \Rightarrow \alpha \rightarrow \sigma} \ (\rightarrow\text{order}).$$

By induction hypothesis, we have $\text{NL} \vdash f(\beta) \Rightarrow f(\alpha)$ and $\text{NL} \vdash f(\sigma) \Rightarrow f(\pi)$. Then, we obtain the required fact:

$$\frac{\vdots \qquad \vdots}{f(\sigma) \Rightarrow f(\pi) \quad f(\beta) \Rightarrow f(\alpha)} \; (\leftarrow\text{order})$$
$$\frac{f(\sigma) \Rightarrow f(\pi) \quad f(\beta) \Rightarrow f(\alpha)}{f(\sigma)\leftarrow f(\alpha) \Rightarrow f(\pi)\leftarrow f(\beta)} \; (\leftarrow\text{order})$$

where $f(\sigma)\leftarrow f(\alpha)$ and $f(\pi)\leftarrow f(\beta)$ coincide with $f(\alpha{\rightarrow}\sigma)$ and $f(\beta{\rightarrow}\pi)$, respectively, by the definition of f.

2. Case ($\sim\!\rightarrow$right): The last inference of P is of the form:

$$\frac{\gamma \Rightarrow \alpha \quad \gamma \Rightarrow \sim\!\beta}{\gamma \Rightarrow \sim(\alpha{\rightarrow}\beta)} \; (\sim\!\rightarrow\text{right}).$$

By induction hypothesis, we have $\text{NL} \vdash f(\alpha) \Rightarrow f(\gamma)$ and $\text{NL} \vdash f(\sim\!\beta) \Rightarrow f(\gamma)$ where $f(\sim\!\beta)$ coincides with $\sim\! f(\beta)$ by the definition of f. Then, we obtain the required fact:

$$\frac{\vdots \qquad \vdots}{\sim\! f(\beta) \Rightarrow f(\gamma) \quad f(\alpha) \Rightarrow f(\gamma)} \; (\sim\!\rightarrow\text{left})$$
$$\frac{\sim\! f(\beta) \Rightarrow f(\gamma) \quad f(\alpha) \Rightarrow f(\gamma)}{\sim(f(\beta)\leftarrow f(\alpha)) \Rightarrow f(\gamma)} \; (\sim\!\rightarrow\text{left})$$

where $\sim(f(\beta)\leftarrow f(\alpha))$ coincides with $f(\sim(\alpha{\rightarrow}\beta))$ by the definition of f.

3. Case ($\sim\!\rightarrow$left2): The last inference of P is of the form:

$$\frac{\sim\!\beta \Rightarrow \delta}{\sim(\alpha{\rightarrow}\beta) \Rightarrow \delta} \; (\sim\!\rightarrow\text{left2}).$$

By induction hypothesis, we have $\text{NL} \vdash f(\delta) \Rightarrow f(\sim\!\beta)$ where $f(\sim\!\beta)$ coincides with $\sim\! f(\beta)$ by the definition of f. Then, we obtain the required fact:

$$\frac{\vdots}{f(\delta) \Rightarrow \sim\! f(\beta)} \; (\sim\!\leftarrow\text{right1})$$
$$\frac{f(\delta) \Rightarrow \sim\! f(\beta)}{f(\delta) \Rightarrow \sim(f(\beta)\leftarrow f(\alpha))} \; (\sim\!\leftarrow\text{right1})$$

where $\sim(f(\beta)\leftarrow f(\alpha))$ coincides with $f(\sim(\alpha{\rightarrow}\beta))$ by the definition of f.

4. Case ($\neg$contraposition): The last inference of P is of the form:

$$\frac{\beta \Rightarrow \alpha}{\neg\alpha \Rightarrow \neg\beta} \; (\neg\text{contraposition}).$$

By induction hypothesis, we have $NL \vdash f(\alpha) \Rightarrow f(\beta)$. Then, we obtain the required fact:

$$\frac{\vdots \\ f(\alpha) \Rightarrow f(\beta)}{\neg' f(\beta) \Rightarrow \neg' f(\alpha)} \ (\neg' \text{contraposition})$$

where $\neg' f(\alpha)$ and $\neg' f(\beta)$ coincide with $f(\neg\alpha)$ and $f(\neg\beta)$, respectively, by the definition of f.

5. Case (cut): The last inference of P is of the form:

$$\frac{\gamma \Rightarrow \alpha \quad \alpha \Rightarrow \delta}{\gamma \Rightarrow \delta} \ (\text{cut}).$$

By induction hypothesis, we have $NL \vdash f(\alpha) \Rightarrow f(\gamma)$ and $NL \vdash f(\delta) \Rightarrow f(\alpha)$. Then, we obtain the required fact:

$$\frac{\overset{\vdots}{f(\delta) \Rightarrow f(\alpha)} \quad \overset{\vdots}{f(\alpha) \Rightarrow f(\gamma)}}{f(\delta) \Rightarrow f(\gamma)} \ (\text{cut}).$$

∎

3 Cut-elimination and its applications

We can show the following cut-elimination theorem via the standard method by Gentzen, which is based on the double induction on grade and rank. But, we illustrate here a simplified proof which was presented in [19] on the basis of the induction on the hight of proof, i.e., the double induction is not needed.

Theorem 3.1 (Cut-elimination for NL). *The rule* (cut) *is admissible in cut-free* NL.

Proof. Consider the following proof P such that there is no occurrence of (cut) in the subproofs P_1 and P_2 of P:

$$\frac{\overset{\vdots \ P_1}{\gamma \Rightarrow \alpha} \quad \overset{\vdots \ P_2}{\alpha \Rightarrow \delta}}{\gamma \Rightarrow \delta} \ (\text{cut}).$$

Then, it is sufficient to show that (cut) in P can be eliminated. This can be proved by induction on the hight of P. It is remarked that we do not need to use the standard double induction on grade and rank. We distinguish the cases according to the last inferences of P_1 and P_2. We show some cases.

1. Case ($\rightarrow$order) ($\rightarrow$left): The last inferences of P_1 and P_2 are ($\rightarrow$order) and ($\rightarrow$left), respectively.

$$
\cfrac{\cfrac{\vdots \qquad \vdots}{\cfrac{\alpha \Rightarrow \gamma_1 \quad \gamma_2 \Rightarrow \beta}{\gamma_1 \rightarrow \gamma_2 \Rightarrow \alpha \rightarrow \beta} \text{ ($\rightarrow$order)} \quad \cfrac{\vdots \qquad \vdots}{\cfrac{\Rightarrow \alpha \quad \beta \Rightarrow \delta}{\alpha \rightarrow \beta \Rightarrow \delta} \text{ ($\rightarrow$left)}}}}{\gamma_1 \rightarrow \gamma_2 \Rightarrow \delta} \text{ (cut)}.
$$

In this case, we obtain the following proof:

$$
\cfrac{\cfrac{\cfrac{\vdots \quad \vdots}{\Rightarrow \alpha \quad \alpha \Rightarrow \gamma_1}}{\Rightarrow \gamma_1} \text{ (cut)} \quad \cfrac{\cfrac{\vdots \quad \vdots}{\gamma_2 \Rightarrow \beta \quad \beta \Rightarrow \delta}}{\gamma_2 \Rightarrow \delta} \text{ (cut)}}{\gamma_1 \rightarrow \gamma_2 \Rightarrow \delta} \text{ ($\rightarrow$left)}
$$

where (cut) in this proof can be eliminated by induction hypothesis.

2. Case ($\sim\rightarrow$right) ($\sim\rightarrow$left2): The last inferences of P_1 and P_2 are ($\sim\rightarrow$right) and ($\sim\rightarrow$left2), respectively.

$$
\cfrac{\cfrac{\cfrac{\vdots \qquad \vdots}{\gamma \Rightarrow \alpha \quad \gamma \Rightarrow \sim\beta}}{\gamma \Rightarrow \sim(\alpha \rightarrow \beta)} \text{ ($\sim\rightarrow$right)} \quad \cfrac{\cfrac{\vdots}{\sim\beta \Rightarrow \delta}}{\sim(\alpha \rightarrow \beta) \Rightarrow \delta} \text{ ($\sim\rightarrow$left2)}}{\gamma \Rightarrow \delta} \text{ (cut)}.
$$

In this case, we obtain the following proof:

$$
\cfrac{\cfrac{\vdots \qquad \vdots}{\gamma \Rightarrow \sim\beta \quad \sim\beta \Rightarrow \delta}}{\gamma \Rightarrow \delta} \text{ (cut)}
$$

where (cut) in this proof can be eliminated by induction hypothesis.

3. Case ($\sim\leftarrow$right1) ($\sim\leftarrow$left): The last inferences of P_1 and P_2 are ($\sim\leftarrow$right1) and ($\sim\leftarrow$left), respectively.

$$
\cfrac{\cfrac{\cfrac{\vdots}{\gamma \Rightarrow \sim\alpha}}{\gamma \Rightarrow \sim(\alpha \leftarrow \beta)} \text{ ($\sim\leftarrow$right1)} \quad \cfrac{\cfrac{\vdots \qquad \vdots}{\sim\alpha \Rightarrow \delta \quad \beta \Rightarrow \delta}}{\sim(\alpha \leftarrow \beta) \Rightarrow \delta} \text{ ($\sim\leftarrow$left)}}{\gamma \Rightarrow \delta} \text{ (cut)}.
$$

In this case, we obtain the following proof:

$$\cfrac{\gamma \Rightarrow \sim\alpha \quad \sim\alpha \Rightarrow \delta}{\gamma \Rightarrow \delta} \ \text{(cut)}$$

where (cut) in this proof can be eliminated by induction hypothesis.

4. Case ($\sim\neg$right) ($\sim\neg$left): The last inferences of P_1 and P_2 are ($\sim\neg$right) and ($\sim\neg$left), respectively.

$$\cfrac{\cfrac{\gamma \Rightarrow \alpha}{\gamma \Rightarrow \sim\neg\alpha} \ (\sim\neg\text{right}) \quad \cfrac{\alpha \Rightarrow \delta}{\sim\neg\alpha \Rightarrow \delta} \ (\sim\neg\text{left})}{\gamma \Rightarrow \delta} \ \text{(cut)}.$$

In this case, we obtain the following proof:

$$\cfrac{\gamma \Rightarrow \alpha \quad \alpha \Rightarrow \delta}{\gamma \Rightarrow \delta} \ \text{(cut)}$$

where (cut) in this proof can be eliminated by induction hypothesis. ∎

Using Theorem 3.1, we can show the decidability of NL.

Theorem 3.2 (Decidability for NL). NL *is decidable.*

Proof. Similar to the proof of the decidability theorem presented in [19] for the single-antecedent/succedent sequent calculus FPQL for the first-order paraconsistent quantum logic. We need the method proposed by Mey [26] for a first-order logic without contraction rules. In this proof, Theorem 3.1 is also used. ∎

Using Theorem 3.1, we can also show the paraconsistency of NL with respect to $\neg$, $\neg'$ and $\sim$. Prior to show this property, we define it as follows.

Definition 3.3. *Let $\sharp$ be a negation connective. A sequent calculus L is called* explosive *with respect to $\sharp$ if for any formulas α and β, the sequent $\alpha \wedge \sharp\alpha \Rightarrow \beta$ is provable in L. It is called* paraconsistent *with respect to $\sharp$ if it is not explosive with respect to $\sharp$.*

Theorem 3.4 (Paraconsistency for NL). *Let $\sharp$ be $\neg$, $\neg'$ or $\sim$. Then, NL is paraconsistent with respect to $\sharp$.*

Proof. Consider a sequent $p \wedge \sharp p \Rightarrow q$ where p and q are distinct propositional variables. Then, the unprovability of this sequent is guaranteed by Theorem 3.1, since there is no cut-free proof of it in NL. ∎

Using Theorem 3.1, we can also show the following characteristic properties of NL.

Theorem 3.5 (Constructive properties for NL). *We have:*

1. *(Conjunction property): If* $\mathrm{NL} \vdash \alpha \wedge \beta \Rightarrow$, *then* $\mathrm{NL} \vdash \alpha \Rightarrow$ *or* $\mathrm{NL} \vdash \beta \Rightarrow$,

2. *(Disjunction property): If* $\mathrm{NL} \vdash \Rightarrow \alpha \vee \beta$, *then* $\mathrm{NL} \vdash \Rightarrow \alpha$ *or* $\mathrm{NL} \vdash \Rightarrow \beta$,

3. *(Universal property): If* $\mathrm{NL} \vdash \forall x \alpha \Rightarrow$, *then* $\mathrm{NL} \vdash \alpha[t/x] \Rightarrow$ *for some term* t,

4. *(Existence property): If* $\mathrm{NL} \vdash \Rightarrow \exists x \alpha$, *then* $\mathrm{NL} \vdash \Rightarrow \alpha[t/x]$ *for some term* t.

Proof. We show only the case (1) below, since the cases (2), (3) and (4) can be obtained in a similar way. Suppose $\vdash \alpha \wedge \beta \Rightarrow$. Consider any cut-free proofs P of $\alpha \wedge \beta \Rightarrow$. We can do so because we have Theorem 3.1. Then, the last inference in P is either ($\wedge$left1) or ($\wedge$left2), and hence, the upper sequent is $\alpha \Rightarrow$ or $\beta \Rightarrow$. Therefore, we have: $\vdash \alpha \Rightarrow$ or $\vdash \beta \Rightarrow$. ∎

Theorem 3.6 (Constructible falsity properties for NL). *Let* $\sharp$ *be* $\neg$, $\neg'$ *or* $\sim$. *Then, we have:*

1. *(Constructible falsity w.r.t. negated conjunction): If* $\mathrm{NL} \vdash \Rightarrow \sharp(\alpha \wedge \beta)$, *then* $\mathrm{NL} \vdash \Rightarrow \sharp \alpha$ *or* $\mathrm{NL} \vdash \Rightarrow \sharp \beta$,

2. *(Constructible falsity w.r.t. negated disjunction) If* $\mathrm{NL} \vdash \sharp(\alpha \vee \beta) \Rightarrow$, *then* $\mathrm{NL} \vdash \sharp \alpha \Rightarrow$ *or* $\mathrm{NL} \vdash \sharp \beta \Rightarrow$,

3. *(Constructible falsity w.r.t. negated universal quantifier): If* $\mathrm{NL} \vdash \Rightarrow \sharp \forall x \alpha$, *then* $\mathrm{NL} \vdash \Rightarrow \sharp \alpha[t/x]$ *for some term* t,

4. *(Constructible falsity w.r.t. negated existential quantifier) If* $\mathrm{NL} \vdash \sharp \exists x \alpha \Rightarrow$, *then* $\mathrm{NL} \vdash \sharp \alpha[t/x] \Rightarrow$ *for some term* t.

Proof. We show only the case (1) below, since the cases (2), (3) and (4) can be obtained in a similar way. The proof of the case $\sharp \equiv \sim$ can straightforwardly be obtained from Theorem 3.1 in a similar manner in the proof of Theorem 3.5. Thus, the proof of the case $\sharp \equiv \neg$ is only shown below. The case $\sharp \equiv \neg'$ can be shown in the same way as that of the case $\sharp \equiv \neg$. Suppose $\vdash \Rightarrow \neg(\alpha \wedge \beta)$. Then, by Theorem 3.1, we have $\vdash \alpha \wedge \beta \Rightarrow$. By Theorem 3.5, we obtain: $\vdash \alpha \Rightarrow$ or $\vdash \beta \Rightarrow$. Therefore, by applying ($\neg$right) to these sequents, we obtain the required fact: $\vdash \Rightarrow \neg\alpha$ or $\vdash \Rightarrow \neg\beta$. ∎

Next, we show the Craig interpolation theorem for NL. Prior to show this theorem, we need to show the following lemma which is a generalized form of the theorem. An expression $V(\alpha)$ for any formula α is used to denote the set of all predicate symbols appearing in α.

Lemma 3.7. *If* NL $\vdash \gamma \Rightarrow \delta$ *where either* γ *or* δ *can be the empty set, then there exists a formula* π *such that*

1. NL $\vdash \gamma \Rightarrow \pi$ *and* NL $\vdash \pi \Rightarrow \delta$,

2. $V(\pi) \subseteq V(\gamma) \cap V(\delta)$.

Proof. In the following, by Theorem 3.1, we consider only cut-free proofs in NL. It is remarked that we need Theorem 3.1 to show this lemma, since the (cut) case in the following proof cannot be shown. Thus, we prove this lemma by induction on the cut-free proofs P of $\gamma \Rightarrow \delta$ in NL. We distinguish the cases according to the last inference of P, and show some cases.

1. Case ($\neg$left): The last inference of P is of the form:

$$\frac{\Rightarrow \alpha}{\neg\alpha \Rightarrow} \; (\neg\text{left}).$$

By induction hypothesis, we have: there exists a formula π such that

(a) $\vdash \Rightarrow \pi$ and $\vdash \pi \Rightarrow \alpha$,

(b) $V(\pi) \subseteq V(\alpha) \cap V(\emptyset) = \emptyset$, i.e., π is $\top$.

Then we obtain:

$$\frac{\vdots \; ind.hyp.}{\frac{\Rightarrow \pi}{\neg\pi \Rightarrow} \; (\neg\text{left})} \qquad \frac{\vdots \; ind.hyp.}{\frac{\pi \Rightarrow \alpha}{\neg\alpha \Rightarrow \neg\pi} \; (\neg\text{contraposition})}.$$

Therefore, we obtain the required fact: there exists a formula $\neg\pi$ such that

(a) $\vdash \neg\alpha \Rightarrow \neg\pi$ and $\vdash \neg\pi \Rightarrow$,

(b) $V(\neg\pi) \subseteq V(\neg\alpha) \cap V(\emptyset) = \emptyset$.

2. Case ($\neg$contraposition): The last inference of P is of the form:

$$\frac{\beta \Rightarrow \alpha}{\neg\alpha \Rightarrow \neg\beta} \ (\neg\text{contraposition}).$$

By induction hypothesis, we have: there exists a formula π such that

(a) $\vdash \beta \Rightarrow \pi$ and $\vdash \pi \Rightarrow \alpha$,

(b) $V(\pi) \subseteq V(\alpha) \cap V(\beta)$.

Then we obtain:

$$\frac{\begin{array}{c} \vdots \ ind.hyp. \\ \beta \Rightarrow \pi \end{array}}{\neg\pi \Rightarrow \neg\beta} \ (\neg\text{contraposition}) \qquad \frac{\begin{array}{c} \vdots \ ind.hyp. \\ \pi \Rightarrow \alpha \end{array}}{\neg\alpha \Rightarrow \neg\pi} \ (\neg\text{contraposition}).$$

Therefore, we obtain the required fact: there exists a formula $\neg\pi$ such that

(a) $\vdash \neg\alpha \Rightarrow \neg\pi$ and $\vdash \neg\pi \Rightarrow \neg\beta$,

(b) $V(\neg\pi) \subseteq V(\neg\alpha) \cap V(\neg\beta)$.

3. Case ($\leftarrow$left): The last inference of P is of the form:

$$\frac{\alpha \Rightarrow \beta}{\alpha{\leftarrow}\beta \Rightarrow} \ (\leftarrow\text{left}).$$

We obtain:

$$\frac{\dfrac{\begin{array}{c} \vdots \ hyp. \\ \alpha \Rightarrow \beta \end{array}}{\alpha{\leftarrow}\beta \Rightarrow}}{\alpha{\leftarrow}\beta \Rightarrow \bot} \ (\text{we-right}).$$

Therefore, we obtain the required fact: there exists the formula $\bot$ such that

(a) $\vdash \alpha{\leftarrow}\beta \Rightarrow \bot$ and $\vdash \bot \Rightarrow$,

(b) $V(\bot) \subseteq V(\alpha{\leftarrow}\beta) \cap V(\emptyset) = \emptyset$.

4. Case ($\leftarrow$right): The last inference of P is of the form:

$$\frac{\gamma \Rightarrow \delta_1 \quad \delta_2 \Rightarrow}{\gamma \Rightarrow \delta_1{\leftarrow}\delta_2} \ (\leftarrow\text{right}).$$

By induction hypothesis, we have: there exists a formula π such that

(a) $\vdash \gamma \Rightarrow \pi$ and $\vdash \pi \Rightarrow \delta_1$,

(b) $V(\pi) \subseteq V(\gamma) \cap V(\delta_1)$.

Then we obtain:

$$\cfrac{\begin{array}{cc} \vdots \; ind.hyp. & \vdots \; hyp. \\ \pi \Rightarrow \delta_1 & \delta_2 \Rightarrow \end{array}}{\pi \Rightarrow \delta_1{\leftarrow}\delta_2} \; (\leftarrow\text{right}).$$

Therefore, we obtain the required fact: there exists a formula π such that

(a) $\vdash \gamma \Rightarrow \pi$ and $\vdash \pi \Rightarrow \delta_1{\leftarrow}\delta_2$,

(b) $V(\pi) \subseteq V(\gamma) \cap V(\delta_1{\leftarrow}\delta_2)$.

5. Case ($\leftarrow$order): The last inference of P is of the form:

$$\cfrac{\sigma \Rightarrow \psi \quad \beta \Rightarrow \alpha}{\sigma{\leftarrow}\alpha \Rightarrow \psi{\leftarrow}\beta} \; (\leftarrow\text{order}).$$

By induction hypothesis, we have:

(a) there exists a formula π_1 such that

  i. $\vdash \sigma \Rightarrow \pi_1$ and $\vdash \pi_1 \Rightarrow \psi$,

  ii. $V(\pi_1) \subseteq V(\sigma) \cap V(\psi)$,

(b) there exists a formula π_2 such that

  i. $\vdash \beta \Rightarrow \pi_2$ and $\vdash \pi_2 \Rightarrow \alpha$,

  ii. $V(\pi_2) \subseteq V(\beta) \cap V(\alpha)$.

Then we obtain:

$$\cfrac{\begin{array}{cc} \vdots \; ind.hyp. & \vdots \; ind.hyp. \\ \sigma \Rightarrow \pi_1 & \pi_2 \Rightarrow \alpha \end{array}}{\sigma{\leftarrow}\alpha \Rightarrow \pi_1{\leftarrow}\pi_2} \; (\leftarrow\text{order}) \qquad \cfrac{\begin{array}{cc} \vdots \; ind.hyp. & \vdots \; ind.hyp. \\ \pi_1 \Rightarrow \psi & \beta \Rightarrow \pi_2 \end{array}}{\pi_1{\leftarrow}\pi_2 \Rightarrow \psi{\leftarrow}\beta} \; (\leftarrow\text{order}).$$

By induction hypothesis, we also obtain:

$$\begin{aligned} V(\pi_1{\leftarrow}\pi_2) &= V(\pi_1) \cup V(\pi_2) \\ &\subseteq (V(\sigma) \cap V(\psi)) \cup (V(\beta) \cap V(\alpha)) \text{ (induction hypothesis)} \\ &\subseteq ((V(\sigma) \cup V(\alpha)) \cap V(\psi)) \cup ((V(\sigma) \cup V(\alpha)) \cap V(\beta)) \\ &= (V(\sigma) \cup V(\alpha)) \cap (V(\psi) \cup V(\beta)) = V(\sigma{\leftarrow}\alpha) \cap V(\psi{\leftarrow}\beta). \end{aligned}$$

Therefore, we have the required fact: there exists a formula $\pi_1{\leftarrow}\pi_2$ such that

(a) $\vdash \sigma \leftarrow \alpha \Rightarrow \pi_1 \leftarrow \pi_2$ and $\vdash \pi_1 \leftarrow \pi_2 \Rightarrow \psi \leftarrow \beta$,

(b) $V(\pi_1 \leftarrow \pi_2) \subseteq V(\sigma \leftarrow \alpha) \cap V(\psi \leftarrow \beta)$.

6. Case ($\sim \rightarrow$left1): The last inference of P is of the form:

$$\frac{\alpha \Rightarrow \delta}{\sim(\alpha \rightarrow \beta) \Rightarrow \delta} \ (\sim \rightarrow\text{left1}).$$

By induction hypothesis, we have: there exists a formula π such that

(a) $\vdash \alpha \Rightarrow \pi$ and $\vdash \pi \Rightarrow \delta$,

(b) $V(\pi) \subseteq V(\alpha) \cap V(\delta)$.

Then we obtain:

$$\frac{\genfrac{}{}{0pt}{}{\vdots \ ind.hyp.}{\alpha \Rightarrow \pi}}{\sim(\alpha \rightarrow \beta) \Rightarrow \pi} \ (\sim \rightarrow\text{left1}).$$

Therefore, we obtain the required fact: there exists a formula π such that

(a) $\vdash \sim(\alpha \rightarrow \beta) \Rightarrow \pi$ and $\vdash \pi \Rightarrow \delta$,

(b) $V(\pi) \subseteq V(\sim(\alpha \rightarrow \beta)) \cap V(\delta)$.

7. Case ($\sim \rightarrow$right): The last inference of P is of the form:

$$\frac{\gamma \Rightarrow \alpha \quad \gamma \Rightarrow \sim\beta}{\gamma \Rightarrow \sim(\alpha \rightarrow \beta)} \ (\sim \rightarrow\text{right}).$$

By induction hypothesis, we have:

(a) there exists a formula π_1 such that

    i. $\vdash \gamma \Rightarrow \pi_1$ and $\vdash \pi_1 \Rightarrow \alpha$,

    ii. $V(\pi_1) \subseteq V(\gamma) \cap V(\alpha)$.

(b) there exists a formula π_2 such that

    i. $\vdash \gamma \Rightarrow \pi_2$ and $\vdash \pi_2 \Rightarrow \sim\beta$,

    ii. $V(\pi_2) \subseteq V(\gamma) \cap V(\sim\beta)$.

Then we obtain:

$$\frac{\genfrac{}{}{0pt}{}{\vdots \ ind.hyp.}{\gamma \Rightarrow \pi_1} \quad \genfrac{}{}{0pt}{}{\vdots \ ind.hyp.}{\gamma \Rightarrow \pi_2}}{\gamma \Rightarrow \pi_1 \wedge \pi_2} \ (\wedge\text{right})$$

$$\cfrac{\cfrac{\vdots\ ind.hyp.}{\cfrac{\pi_1 \Rightarrow \alpha}{\pi_1 \wedge \pi_2 \Rightarrow \alpha}\ (\wedge\text{left2})} \quad \cfrac{\cfrac{\vdots\ ind.hyp.}{\pi_2 \Rightarrow \sim\beta}}{\pi_1 \wedge \pi_2 \Rightarrow \sim\beta}\ (\wedge\text{left1})}{\pi_1 \wedge \pi_2 \Rightarrow \sim(\alpha{\to}\beta)}\ (\sim{\to}\text{right}).$$

Therefore, we obtain the required fact: there exists a formula $\pi_1 \wedge \pi_2$ such that

(a) $\vdash \gamma \Rightarrow \pi_1 \wedge \pi_2$ and $\vdash \pi_1 \wedge \pi_2 \Rightarrow \sim(\alpha{\to}\beta)$,

(b) $V(\pi_1 \wedge \pi_2) \subseteq V(\gamma) \cap V(\sim(\alpha{\to}\beta))$. ∎

Theorem 3.8 (Craig interpolation for NL). *If* NL $\vdash \alpha \Rightarrow \beta$ *for any formulas α and β, then there exists a formula π such that*

1. NL $\vdash \alpha \Rightarrow \pi$ *and* NL $\vdash \pi \Rightarrow \beta$,

2. $V(\pi) \subseteq V(\alpha) \cap V(\beta)$.

Proof. This theorem is obtained as a special case of Lemma 3.7. ∎

4 Uniform provability

To obtain a cut-free goal-directed proof called the *uniform proof* [27], we have to consider the following conditions:

1. the left (right) introduction rules are permuted upward (downward) as possible in a given cut-free proof,

2. the formulas we can deal with are restricted to a kind of hereditary Harrop formulas.

In a cut-free proof, some inference rules such as ($\vee$left) in NL cannot be permuted upward (over some right introduction rules), i.e., the condition (1) does not hold for a usual unrestricted formula. We thus have to impose the condition (2) in order to delete such left introduction rules in the cut-free proof.

As mentioned, the order of the inference rules in a cut-free proof of a sequent system can often be permuted upward or downward. This property is called the *inference permutability* [23]. In an inference rule, the formulas which are present in the premise(s) but not in the conclusion are called the *active formulas*, and the formula(s) which is(are) present in the conclusion but not in the premise(s) is(are) called the *principal formula(s)*. For example, in the rule:

$$\frac{\alpha \Rightarrow \beta \quad \pi \Rightarrow \sigma}{\beta{\to}\pi \Rightarrow \alpha{\to}\sigma}\ (\to\text{order})$$

where $\beta{\rightarrow}\pi$ and $\alpha{\rightarrow}\sigma$ are the principal formulas of ($\rightarrow$order), and α, β, π and σ are active formulas of ($\rightarrow$order).

When looking to permute the order of two inference rules in a proof, it is necessary to check that the principal formula of the upper inference rule is not an active formula of the lower one; otherwise, no permutation is possible. When this fact occurs, the two inference rules are said to be in *permutation position*. For example, consider the proof of the form:

$$\frac{\quad \vdots \quad \cfrac{\beta \Rightarrow {\sim}\pi}{\beta \Rightarrow {\sim}(\pi \wedge \sigma)}\,({\sim}\wedge\text{right1})}{\alpha{\rightarrow}\beta \Rightarrow {\sim}(\pi \wedge \sigma)}\,(\rightarrow\text{left})$$

where the active formula of (${\sim}\wedge$right1) is ${\sim}\pi$, the principal formula of (${\sim}\wedge$right1) is ${\sim}(\pi \wedge \sigma)$, the active formulas of ($\rightarrow$left) are α and β, and the principal formula of ($\rightarrow$left) is $\alpha{\rightarrow}\beta$. In this case, the rules (${\sim}\wedge$right1) and ($\rightarrow$left) are in permutation position, and then ($\rightarrow$left) can be permuted upward:

$$\frac{\cfrac{\Rightarrow\alpha \quad \beta \Rightarrow {\sim}\pi}{\alpha{\rightarrow}\beta \Rightarrow {\sim}\pi}\,(\rightarrow\text{left})}{\alpha{\rightarrow}\beta \Rightarrow {\sim}(\pi \wedge \sigma)}\,({\sim}\wedge\text{right1}).$$

Theorem 4.1 (Inference permutability for NL). *For a cut-free proof P of NL, an inference rule $I \in \{$(we-left), (we-right), ($\wedge$left1), ($\wedge$left2), ($\rightarrow$left), ($\leftarrow$right), ($\forall$left), (${\sim}{\sim}$left), (${\sim}\vee$left1), (${\sim}\vee$left2), (${\sim}{\rightarrow}$left1), (${\sim}{\rightarrow}$left2), (${\sim}\neg$left), (${\sim}\neg'$left), (${\sim}\exists$left)$\}$ appearing in P can be permuted upward as possible, i.e., up to initial sequent(s) or a position that the active formula(s) of I is(are) introduced by other applications of inference rules.*

Proof. This theorem can be proved in a similar way as for LJ. We consider only the case for ($\leftarrow$right), since other cases can also be shown in a similar way. In this case, we have to show that for every inference rule I just over ($\leftarrow$right) in a given cut-free proof, ($\leftarrow$right) can be permuted over I. We show only the following case:

$$\frac{\cfrac{\Rightarrow\alpha_1 \quad \beta_1 \Rightarrow \alpha_2}{\alpha_1{\rightarrow}\beta_1 \Rightarrow \alpha_2}\,(\rightarrow\text{left}) \qquad \vdots\;\beta_2 \Rightarrow}{\alpha_1{\rightarrow}\beta_1 \Rightarrow \alpha_2{\leftarrow}\beta_2}\,(\leftarrow\text{right}).$$

This can be transformed into the following proof:

$$\cfrac{\cfrac{\vdots}{\Rightarrow \alpha_1} \quad \cfrac{\cfrac{\vdots}{\beta_1 \Rightarrow \alpha_2} \quad \cfrac{\vdots}{\beta_2 \Rightarrow}}{\beta_1 \Rightarrow \alpha_2 \leftarrow \beta_2} (\leftarrow\text{right})}{\alpha_1 \rightarrow \beta_1 \Rightarrow \alpha_2 \leftarrow \beta_2} (\rightarrow\text{left}).$$

∎

The logic programming interpretation of a sequent $\gamma \Rightarrow \delta$ is that the antecedent γ represents the program and the succedent δ represents the goal. When searching for a proof of $\gamma \Rightarrow \delta$ (i.e., performing computation), the search should be "driven" by δ, i.e., "goal-directed." The proof-theoretic characterization of this property is given by the notion of uniform proofs [27]. A uniform proof for NL is a cut-free proof (in NL) in which each occurrence of a sequent whose succedent contains a non-atomic (or non-$\sim$-negated-atomic) formula is the lower sequent of the inference rule that introduces its outermost connective (or $\sim$-negated-outermost connective).

Definition 4.2 (Uniform proof for NL). *A uniform proof for NL is a cut-free proof in NL such that, for each occurrence of a sequent $\gamma \Rightarrow \delta$ in it, the following conditions are satisfied.*

1. *If δ is $\alpha \wedge \beta$, then that sequent is inferred by ($\wedge$right) from $\gamma \Rightarrow \alpha$ and $\gamma \Rightarrow \beta$.*

2. *If δ is $\alpha \vee \beta$, then that sequent is inferred by ($\vee$right1) or ($\vee$right2) from $\gamma \Rightarrow \alpha$ or $\gamma \Rightarrow \beta$, respectively.*

3. *If δ is $\alpha \rightarrow \beta$ and γ is the empty set, then that sequent is inferred by ($\rightarrow$right) from $\alpha \Rightarrow \beta$.*

4. *If δ is $\alpha \rightarrow \sigma$ and γ is $\beta \rightarrow \pi$, then that sequent is inferred by ($\rightarrow$order) from $\alpha \Rightarrow \beta$ and $\pi \Rightarrow \sigma$.*

5. *If δ is $\alpha \leftarrow \beta$, then that sequent is inferred by ($\leftarrow$right) from $\gamma \Rightarrow \alpha$ and $\beta \Rightarrow$.*

6. *If δ is $\pi \leftarrow \beta$ and γ is $\sigma \rightarrow \alpha$, then that sequent is inferred by ($\leftarrow$order) from $\sigma \Rightarrow \pi$ and $\beta \Rightarrow \alpha$.*

7. *If δ is $\neg\alpha$ and γ is the empty set, then that sequent is inferred by ($\neg$right) from $\alpha \Rightarrow$.*

8. *If δ is $\neg\beta$ and γ is $\neg\alpha$, then that sequent is inferred by ($\neg$contraposition) from $\beta \Rightarrow \alpha$.*

9. If δ is $\neg'\alpha$ and γ is the empty set, then that sequent is inferred by ($\neg'$right) from $\alpha \Rightarrow$.

10. If δ is $\neg'\beta$ and γ is $\neg'\alpha$, then that sequent is inferred by ($\neg'$contraposition) from $\beta \Rightarrow \alpha$.

11. If δ is $\forall x\alpha$, then that sequent is inferred by ($\forall$right) from $\gamma \Rightarrow \alpha[a/x]$ where a is a free variable which does not occur in $\gamma \Rightarrow \forall x\alpha$.

12. If δ is $\exists x\alpha$, then that sequent is inferred by ($\exists$right) from $\gamma \Rightarrow \alpha[t/x]$ where t is an arbitrary term.

13. If δ is $\sim\sim\alpha$, then that sequent is inferred by ($\sim\sim$right) from $\gamma \Rightarrow \alpha$.

14. If δ is $\sim(\alpha \wedge \beta)$, then that sequent is inferred by ($\sim\wedge$right1) or ($\sim\wedge$right2) from $\gamma \Rightarrow \sim\alpha$ or $\gamma \Rightarrow \sim\beta$, respectively.

15. If δ is $\sim(\alpha \vee \beta)$, then that sequent is inferred by ($\sim\vee$right) from $\gamma \Rightarrow \sim\alpha$ and $\gamma \Rightarrow \sim\beta$.

16. If δ is $\sim(\alpha \rightarrow \beta)$, then that sequent is inferred by ($\sim\rightarrow$right) from $\gamma \Rightarrow \alpha$ and $\gamma \Rightarrow \sim\beta$.

17. If δ is $\sim(\alpha \leftarrow \beta)$, then that sequent is inferred by ($\sim\leftarrow$right1) or ($\sim\leftarrow$right2) from $\gamma \Rightarrow \sim\alpha$ or $\gamma \Rightarrow \beta$, respectively.

18. If δ is $\sim\neg\alpha$, then that sequent is inferred by ($\sim\neg$right) from $\gamma \Rightarrow \alpha$.

19. If δ is $\sim\neg'\alpha$, then that sequent is inferred by ($\sim\neg'$right) from $\gamma \Rightarrow \alpha$.

20. If δ is $\sim\forall x\alpha$, then that sequent is inferred by ($\sim\forall$right) from $\gamma \Rightarrow \sim\alpha[t/x]$ where t is an arbitrary term.

21. If δ is $\sim\exists x\alpha$, then that sequent is inferred by ($\sim\exists$right) from $\gamma \Rightarrow \sim\alpha[a/x]$ where a is a free variable which does not occur in $\gamma \Rightarrow \sim\exists x\alpha$.

An expression $L \vdash_U \gamma \Rightarrow \delta$ denotes that there is a uniform proof of the sequent $\gamma \Rightarrow \delta$ in a sequent calculus L. This expression intuitively means that an interpreter L succeeds on the goal δ given the program γ.

Definition 4.3 (Extended first-order hereditary Harrop formulas). Extended first-order hereditary Harrop formulas *are defined by the following grammar using mutual induction, assuming p represents atomic formulas:*

$$\gamma ::= p \mid \gamma \wedge \gamma \mid \delta{\to}p \mid \delta{\to}{\sim}p \mid p{\leftarrow}\delta \mid {\sim}p{\leftarrow}\delta \mid \forall x\gamma \mid$$
$${\sim}p \mid {\sim}{\sim}\gamma \mid {\sim}(\gamma \vee \gamma) \mid {\sim}(\gamma{\to}\gamma) \mid {\sim}\neg\gamma \mid {\sim}\neg'\gamma \mid {\sim}\exists x\gamma.$$
$$\delta ::= p \mid \delta \wedge \delta \mid \delta \vee \delta \mid \gamma{\to}\delta \mid \delta{\leftarrow}\gamma \mid \neg p \mid \neg'p \mid \forall x\delta \mid \exists x\delta \mid {\sim}\delta.$$

The formulas γ which are defined above are called program formulas, *and the formulas δ which are defined above are called* goal formulas.

Theorem 4.4 (Uniform provability for NL). *Let γ be a program-formula and δ be a goal-formula. Then, we have:* $\text{NL} \vdash \gamma \Rightarrow \delta$ *if and only if* $\text{NL} \vdash_U \gamma \Rightarrow \delta$.

Proof. Similar to the proof in [27]. Since any uniform proof in NL is also a cut-free proof in NL, we have that $\text{NL} \vdash_U \gamma \Rightarrow \delta$ implies $\text{NL} \vdash \gamma \Rightarrow \delta$. Thus, we show that $\text{NL} \vdash \gamma \Rightarrow \delta$ implies $\text{NL} \vdash_U \gamma \Rightarrow \delta$. Suppose $\text{NL} \vdash \gamma \Rightarrow \delta$. Then we have $\text{NL} -$ (cut) $\vdash \gamma \Rightarrow \delta$ by the cut-elimination theorem for NL. Let P be a cut-free proof of $\gamma \Rightarrow \delta$ in NL. To obtain a uniform proof, we now consider to lift-up the left-inference rules, (we-left) and (we-right) in P over the right rules. By the cut-elimination theorem, we have the subformula-property-like property for NL, and hence every formula occurring in any sequent in P is a subformula or $\sim$-negated-subformula of some formulas occurring in $\gamma \Rightarrow \delta$. By this fact and Definition 4.3, there is no application of the rule ($\vee$left), ($\neg$left), ($\neg'$left), ($\exists$left), ($\sim\wedge$left), ($\sim\leftarrow$left) or ($\sim\forall$left) in P. By using Theorem 4.1, we then have the fact that the remained structural rules (we-left) and (we-right) and the remained left-inference rules ($\wedge$left1), ($\wedge$left2), ($\to$left), ($\forall$left), ($\sim\sim$left), ($\sim\vee$left1), ($\sim\vee$left2), ($\sim\to$left1), ($\sim\to$left2), ($\sim\neg$left), ($\sim\neg'$left) and ($\sim\exists$left) appearing in P can move the upper direction in P as lift-up as possible, i.e., all of these possible left-rules and weakening rules permute upward as possible. Note that the inference rules ($\leftarrow$left), ($\to$order), ($\leftarrow$order), ($\neg$left), ($\neg'$left), ($\neg$contraposition) and ($\neg'$contraposition) cannot permute upward, since these rules cannot be in a permutation position. Then we can eliminate non-uniform inferences by permuting left-rules up. A resulting proof P' by lifting-up these inference rules appropriately is a uniform proof of $\gamma \Rightarrow \delta$, and hence $\text{NL} \vdash_U \gamma \Rightarrow \delta$. ∎

Different from Gentzen's LJ, we can obtain the dual version of the uniform provability theorem for NL. An expression $L \vdash_D \gamma \Rightarrow \delta$ denotes that there is a *dual uniform proof* of the sequent $\gamma \Rightarrow \delta$ in a sequent calculus L.

Definition 4.5 (Dual extended first-order hereditary Harrop formulas). *Dual extended first-order hereditary Harrop formulas are defined by the following grammar using mutual induction, assuming p represents atomic formulas:*

$$\gamma ::= p \mid \gamma \wedge \gamma \mid \gamma \vee \gamma \mid \delta{\to}\gamma \mid \gamma{\leftarrow}\delta \mid \neg p \mid \neg'p \mid \forall x\gamma \mid \exists x\gamma \mid {\sim}\gamma.$$

81

$$\delta ::= \; p \mid \delta \vee \delta \mid p{\rightarrow}\gamma \mid {\sim}p{\rightarrow}\gamma \mid \gamma{\leftarrow}p \mid \gamma{\leftarrow}{\sim}p \mid \exists x\delta \mid$$
$${\sim}p \mid {\sim}{\sim}\delta \mid {\sim}(\delta \wedge \delta) \mid {\sim}(\delta{\leftarrow}\delta) \mid {\sim}\neg\delta \mid {\sim}\neg'\delta \mid {\sim}\forall x\delta.$$

The formulas γ which are defined above are called dual goal formulas, *and the formulas δ which are defined above are called* dual program formulas.

Theorem 4.6 (Dual uniform provability for NL). *Let γ be a dual-goal-formula and δ be a dual-program-formula. Then, we have:* $\mathrm{NL} \vdash \gamma \Rightarrow \delta$ *if and only if* $\mathrm{NL} \vdash_D \gamma \Rightarrow \delta$.

Proof. By Theorems 2.4 and 4.5. ∎

5 Extension with naive comprehension

In what follows, we consider to extend NL with some naive comprehension rules in the naive set theory. It will be shown that the extended system $\mathrm{NL}_\in$ is consistent, and Russell's (like) paradox does not imply the inconsistency of $\mathrm{NL}_\in$.

Definition 5.1. Terms *and* formulas *of first-order Nelsonian paraconsistent quantum logic with naive comprehension* are simultaneously defined as follows:

1. *variables $x, y, z, ...$ are terms,*

2. *if α is a formula and x is a variable, then $\{x \mid \alpha\}$ is a term,*

3. *if t and u are terms, then $t \in u$ is a formula,*

4. *$\top$ and $\bot$ are formulas,*

5. *if α and β are formulas, then so are $\alpha \wedge \beta$, $\alpha \vee \beta$, $\alpha{\rightarrow}\beta$, $\alpha{\leftarrow}\beta$, $\neg\alpha$, $\neg'\alpha$ and ${\sim}\alpha$,*

6. *if α is a formula and x is a variable, then $\forall x\alpha$ is a formula.*

The same notations and notions as those for NL are also used in the following. Small letters $s, t, u, ...$ are used to denote terms. An expression $u[t/x]$ means the formula which is obtained from the term u by replacing all free occurrences of the variable x in u with the term t, but avoiding a clash of variables by a suitable renaming of bound variables.

A sequent calculus $\mathrm{NL}_\in$ for the first-order Nelsonian paraconsistent quantum logic with naive comprehension is defined below.

Definition 5.2 (NL$_\in$). *In the following, γ and δ in the sequents appearing in the initial sequents and inference rules represent a formula or the empty set.*

NL$_\in$ is obtained from NL by adding the comprehension inference rules of the form:

$$\frac{\alpha[t/x] \Rightarrow \delta}{t \in \{x \mid \alpha\} \Rightarrow \delta} \ (\in \text{left}) \qquad \frac{\gamma \Rightarrow \alpha[t/x]}{\gamma \Rightarrow t \in \{x \mid \alpha\}} \ (\in \text{right})$$

$$\frac{\sim\alpha[t/x] \Rightarrow \delta}{\sim(t \in \{x \mid \alpha\}) \Rightarrow \delta} \ (\sim\in \text{left}) \qquad \frac{\gamma \Rightarrow \sim\alpha[t/x]}{\gamma \Rightarrow \sim(t \in \{x \mid \alpha\})} \ (\sim\in \text{right}).$$

Remark 5.3. *We make the following remarks.*

1. *The following sequents which just correspond to the naive comprehension axiom in the naive set theory are provable in cut-free NL$_\in$: For any formula α,*

   $$\alpha[t/x] \Leftrightarrow t \in \{x \mid \alpha\}.$$

2. *By using the naive comprehension axiom $\alpha[t/x] \leftrightarrow t \in \{x \mid \alpha\}$, Russell's (like) paradox $\alpha \leftrightarrow \sharp\alpha$ is obtained by putting α as $s \in s$ and s as $\{x \mid \sharp(x \in x)\}$ where $\sharp$ is $\neg, \neg'$ or $\sim$.*

3. *However, this paradox does not imply the inconsistency of NL$_\in$, because NL$_\in$ is consistent, i.e., $\Rightarrow$ is not provable in NL$_\in$. This consistency is a consequent of the cut-elimination theorem for NL$_\in$.*

Theorem 5.4 (Cut-elimination for NL$_\in$). *The rule (cut) is admissible in cut-free NL$_\in$.*

Proof. Consider the following proof P such that there is no occurrence of (cut) in the subproofs P_1 and P_2 of P:

$$\frac{\begin{array}{cc} \vdots \ P_1 & \vdots \ P_2 \\ \gamma \Rightarrow \alpha & \alpha \Rightarrow \delta \end{array}}{\gamma \Rightarrow \delta} \ (\text{cut}).$$

Then, it is sufficient to show that (cut) in P can be eliminated. This can be proved by induction on the hight of P. We distinguish the cases according to the last inferences of P_1 and P_2. We show only the following case.

Case ($\sim\in$ right) ($\sim\in$ left): The last inferences of P_1 and P_2 are ($\sim\in$ right) and ($\sim\in$ left), respectively.

$$\frac{\dfrac{\begin{array}{c}\vdots\\\gamma \Rightarrow \sim\alpha[t/x]\end{array}}{\gamma \Rightarrow \sim(t \in \{x \mid \alpha\})} \ (\sim\in \text{right}) \quad \dfrac{\begin{array}{c}\vdots\\\sim\alpha[t/x] \Rightarrow \delta\end{array}}{\sim(t \in \{x \mid \alpha\}) \Rightarrow \delta} \ (\sim\in \text{left})}{\gamma \Rightarrow \delta} \ (\text{cut}).$$

In this case, we obtain the following proof:

$$\frac{\vdots \qquad\qquad \vdots}{\gamma \Rightarrow \sim\!\alpha[t/x] \quad \sim\!\alpha[t/x] \Rightarrow \delta} \atop \gamma \Rightarrow \delta} \text{(cut)}$$

where (cut) in this proof can be eliminated by induction hypothesis. ∎

Remark 5.5. *By using this cut-elimination theorem, we can obtain the conjunction, disjunction, universal, existence and constructible falsity properties for* $\mathrm{NL}_\in$. *We can also show the Craig interpolation and (dual-)uniform provability (in an appropriate setting) theorems for* $\mathrm{NL}_\in$ *in a similar way as for* NL. *However, we have not yet known the (un)decidability of* $\mathrm{NL}_\in$.

6 Conclusions, remarks, and related works

In this paper, a single-antecedent/succedent sequent calculus NL for the first-order Nelsonian paraconsistent quantum logic was introduced for investigating several desirable properties of the logic. The duality and cut-elimination theorems for NL were proved, and some applications of the cut-elimination theorem were obtained. These applications consist of the decidability, the four constructive properties, the four constructible falsity properties, the Craig interpolation property, and the uniform and dual-uniform provabilities. The characteristic properties of NL and the first-order Nelsonian paraconsistent quantum logic are thus the duality, the conjunction and universal properties, the constructible falsity properties with respect to negated disjunction and negated existential quantifiers, and the dual-uniform provability. It is known that these properties do not hold for the standard paraconsistent logics. It was also shown in this paper that NL is useful as a basis for automated theorem proving and logic programming, because NL was shown to be decidable and the uniform and dual-uniform provability theorems were proved. Furthermore, an extension $\mathrm{NL}_\in$ of NL, which is obtained from NL by adding several naive comprehension rules from the naive set theory, was introduced. It was shown that the extended system $\mathrm{NL}_\in$ is consistent, and Russell's (like) paradox does not imply the inconsistency of $\mathrm{NL}_\in$.

We remark that the proposed systems can be naturally extended by adding some modal operators. For example, we can consider extensions MNL and $\mathrm{MNL}_\in$ of NL and $\mathrm{NL}_\in$, respectively, by adding the S4-type modal operators $\square$ (necessity) and $\diamond$ (possibility).

We now define MNL and $\mathrm{MNL}_\in$ as follows.

Definition 6.1 (MNL and $\mathrm{NL}_\in$). *MNL and* $\mathrm{MNL}_\in$ *are obtained from* NL *and* $\mathrm{NL}_\in$, *respectively, by adding the logical inference rules of the form:*

$$\frac{\alpha \Rightarrow \delta}{\Box\alpha \Rightarrow \delta}\ (\Box\text{left}) \qquad \frac{\Box\gamma \Rightarrow \alpha}{\Box\gamma \Rightarrow \Box\alpha}\ (\Box\text{right})$$

$$\frac{\alpha \Rightarrow \Diamond\delta}{\Diamond\alpha \Rightarrow \Diamond\delta}\ (\Diamond\text{left}) \qquad \frac{\gamma \Rightarrow \alpha}{\gamma \Rightarrow \Diamond\alpha}\ (\Diamond\text{right})$$

$$\frac{\sim\alpha \Rightarrow \sim\Box\delta}{\sim\Box\alpha \Rightarrow \sim\Box\delta}\ (\sim\Box\text{left}) \qquad \frac{\gamma \Rightarrow \sim\alpha}{\gamma \Rightarrow \sim\Box\alpha}\ (\sim\Box\text{right})$$

$$\frac{\sim\alpha \Rightarrow \delta}{\sim\Diamond\alpha \Rightarrow \delta}\ (\sim\Diamond\text{left}) \qquad \frac{\sim\Diamond\gamma \Rightarrow \sim\alpha}{\sim\Diamond\gamma \Rightarrow \sim\Diamond\alpha}\ (\sim\Diamond\text{right}).$$

The following sequents are provable in cut-free MNL: For any formula α,

1. $\sim\Box\alpha \Leftrightarrow \Diamond\sim\alpha$,

2. $\sim\Diamond\alpha \Leftrightarrow \Box\sim\alpha$.

We can prove the duality and cut-elimination theorems for MNL and $\mathrm{MNL}_\in$, and by using the cut-elimination theorems, we can obtain the conjunction, disjunction, universal, existence, and constructible falsity properties for MNL and $\mathrm{MNL}_\in$. The duality theorems for MNL and $\mathrm{MNL}_\in$ can be proved using an extended mapping f with the conditions $f(\Box\alpha) := \Diamond f(\alpha)$ and $f(\Diamond\alpha) := \Box f(\alpha)$.

Going forward, we intend to show the decidability of MNL, the Craig interpolation and (dual-)uniform provability (in an appropriate setting) theorems for MNL and $\mathrm{MNL}_\in$, and the (un)decidabilities of $\mathrm{NL}_\in$ and higher-order extensions of NL, $\mathrm{NL}_\in$, MNL, and $\mathrm{MNL}_\in$. Based on these extended systems of NL, we intend to develop logic programming languages (based on the uniform probabilities) and automated theorem proving systems (based on the single-antecedent/succedent sequent calculi).

The background of and motivation for original quantum logic are explained mainly based on [7], as follows. The original quantum logic, which was introduced by Birkhoff and von Neumann [5], was described as a logical abstraction from the algebraic structure of the projectors (or the closed subspaces) in a Hilbert space. According to the standard axiomatization of quantum mechanics, a physical system S (e.g. an electron) is interpreted as a Hilbert space $\mathcal{H}$. Pure states that S may assume in different times are interpreted as vectors ϕ of $\mathcal{H}$. From an intuitive perspective, a state ϕ represents a preparation procedure that is realized by some apparatus that produces individual samples of a physical system under well-defined and repeatable conditions. The closed subspaces X of $\mathcal{H}$ (or the projectors P_X

with a given range X) correspond to the physical properties that S can satisfy. The mathematical interpretation of the physical properties of S provides a quantum logic model, as projectors (or closed subspaces) of $\mathcal{H}$ give us the algebraic semantics for quantum logic.

Projectors are not the only operators from which Born probabilities can be defined. Considering the class $\mathcal{E}(\mathcal{H})$ of all linear bounded operators, $\mathcal{E}(\mathcal{H})$ then includes the set of the projectors and represents the maximal possible notion of a physical property, in agreement with the probabilistic rules of quantum mechanics. In the framework of the operational approach to quantum theory [24, 25], the elements of $\mathcal{E}(\mathcal{H})$ are called *effects*. From an intuitive perspective, effects can be regarded as being produced by yes-no measurement devices that test questions about the physical system under consideration. Projectors can be associated to *sharp* questions having the form "the value for the observable physical quantity A lies in the Borel set B", whereas effects can also represent *fuzzy* (or *vague*) questions having the form "the value of the observable physical quantity A lies in the fuzzy Borel set B." For a systematic presentation of the operational approach to quantum mechanics, see [24, 25].

Next, the background of and motivation for the paraconsistent quantum logic are explained based mainly on [10, 15], as follows. It was argued by Dalla Chiara and Giuntini in [10] that the paraconsistent quantum logic appears to be an interesting logical abstraction from the class of all effects in the operational approach to quantum mechanics. As explained above, effects are, in the operational approach to quantum mechanics, natural generalizations of projectors (any projector is an effect, but not vice versa) [10]. Giuntini and Greuling in [10] affirm that the paraconsistent quantum logic represents a somewhat approximate logical abstraction from the class of all effect-frames. Moreover, from an intuitive point of view, the possible worlds of an effect-frame (with respect to the paraconsistent quantum logic) represent possible physical properties (or descriptions), which are generally incomplete, of state of quantum systems [10]. It was also shown in [15] that some algebraic structures of the set of all effects serve as a model of the paraconsistent quantum logic.

Finally, we illustrate several closely related works on the paraconsistent quantum logic and their related logic as follows. *Fuzzy intuitionistic quantum logic* (or *Brouwer–Zadeh logic*) was introduced by Cattaneo and Nisticó [6]. The paraconsistent quantum logic has turned out to be a type of minimal fuzzy fragment of fuzzy intuitionistic quantum logic. A fuzzy-like paraconsistent negation, which results in paraconsistent behavior, was investigated by Cattaneo, Dalla Chiara, and Giuntini [7], based on fuzzy intuitionistic quantum logic. Algebraic and Kripke-type semantics for fuzzy intuitionistic quantum logic were also investigated in [7]. For some recent topics on quantum logics and quantum computation, see e.g. [11] in which it was explained how Birkhoff and von Neumann's quantum logic, and the more recent forms of fuzzy quantum logic, have naturally emerged from the mathematical environment of quantum theory.

Acknowledgments. This research was supported by JSPS KAKENHI Grant Numbers JP18K11171, JP16KK0007, and JSPS Core-to-Core Program (A. Advanced Research Networks).

References

[1] A. Almukdad and D. Nelson, Constructible falsity and inexact predicates, Journal of Symbolic Logic 49 (1), pp. 231-233, 1984.

[2] N.D. Belnap, A useful four-valued logic, In Modern Uses of Multiple-Valued Logic, G. Epstein and J. M. Dunn, eds., Dordrecht: Reidel, pp. 7-37, 1977.

[3] N.D. Belnap, How computer should think, pp. 30-56, In Contemporary Aspects of Philosophy, (G. Ryle ed.), Oriel Press, Stocksfield, 1977.

[4] J.-Y. Beziau, Monosequent proof systems, in: C. Caleiro, F. Dionisio, P. Gouveia, P. Mateus and J. Rasga (eds.), Logic and Computation – Essays in Honor of Amilcar Sernadas, pp. 111-137, College Publication, London, 2017.

[5] G. Birkhoff and J. von Neumann, The logic of quantum mechanics, Annals of Mathematics 37, pp. 823-843, 1936.

[6] G. Cattaneo and G. Nisticó, Brouwer–Zadeh posets and three-valued Łukasiewicz posets, Fuzzy Sets and Systems 33, pp. 165-190, 1989.

[7] G. Cattaneo, M. L. Dalla Chiara, and R. Giuntini, Fuzzy intuitionistic quantum logic, Studia Logica 52, pp. 419-442, 1993.

[8] W. Craig, Three uses of the Herbrand-Gentzen theorem in relating model theory and proof theory, Journal of Symbolic Logic 22 (3), pp. 269-285, 1957.

[9] J. Czermak, A remark on Gentzen's calculus of sequents, Notre Dame Journal of Formal Logic 18, pp. 471-474, 1977.

[10] M.L. Dalla Chiara and R. Giuntini, Paraconsistent quantum logics, Foundations of Physics 19 (7), pp. 891-904, 1989.

[11] M. L. Dalla Chiara, R. Giuntini, R. Leporini, and G. Sergioli, Quantum computation and logic - How quantum computers have inspired logical investigations, Trends in Logic 48, Springer, 2018.

[12] J.M. Dunn, Intuitive semantics for first-degree entailment and 'coupled trees', Philosophical Studies 29 (3), pp. 149-168, 1976.

[13] C. Faggian and G. Sambin, From basic logic to quantum logics with cut-elimination, International Journal of Theoretical Physics 37 (1), pp. 31-37, 1998.

[14] G. Gentzen, Collected papers of Gerhard Gentzen, M.E. Szabo, ed., Studies in logic and the foundations of mathematics, North-Holland (English translation), 1969.

[15] R. Giuntini and H. Greuling, Toward a formal language for unsharp properties, Foundations of physics 19 (7), pp. 931-945, 1989.

[16] N.D. Goodman, The logic of contradiction, Zeitschrift für Mathematische Logik und Grundlagen der Mathematik 27, pp. 119-126, 1981.

[17] V.N. Grishin, Predicate and set theoretic calculi based on logic without contraction rules (Russian), Izvestiya Akademii Nauk SSSR Seriya Mathematicheskaya 45 (1), pp. 47-68, 1981, English translation in Math. USSR Izv. 18 (1), pp. 41-59, 1982.

[18] N. Kamide, A uniform proof-theoretic foundation for abstract paraconsistent logic programming, Journal of Functional and Logic Programming 2007, pp. 1-36, 2007.

[19] N. Kamide, Proof theory of paraconsistent quantum logic, Journal of Philosophical Logic 47 (2), pp. 301-324, 2018.

[20] N. Kamide, Extending paraconsistent quantum logic: A single-antecedent/ succedent system approach, Mathematical Logic Quarterly 64 (4-5), pp. 371-386, 2018.

[21] N. Kamide, First-order Nelsonian paraconsistent quantum logic, Proceedings of the 49th IEEE International Symposium on Multiple-Valued Logic (ISMVL 2019), pp. 176-181, 2019.

[22] N. Kamide and H. Wansing, Proof theory of N4-related paraconsistent logics, Studies in Logic 54, 401 pages, College Publications, 2015.

[23] S.C. Kleene, Permutability of inferences in Gentzen's calculi LK and LJ, Memoirs of the American Mathematical Society 10, pp. 1-26, 1952.

[24] K. Kraus, States, effects, and operations - Fundamental notions of quantum theory, Lectures in Mathematical Physics at the University of Texas at Austin, Springer, Berlin, New York, 1983.

[25] G. Ludwig, Foundations of quantum mechanics I, Theoretical and Mathematical Physics, Springer, Berlin, New York, 1983.

[26] D. Mey, A predicate calculus with control of derivations, Proceedings of the 3rd Workshop on Computer Science Logic, Lecture Notes in Computer Science 440, pp. 254-266, 1989.

[27] D. Miller, G. Nadathur, F. Pfenning and A. Scedrov, Uniform proofs as a foundation for logic programming, Annals of Pure and Applied Logic 51, pp. 125-157, 1991.

[28] D. Nelson, Constructible falsity, Journal of Symbolic Logic 14, pp. 16-26, 1949.

[29] G. Priest, Paraconsistent logic, Handbook of Philosophical Logic (Second Edition), Vol. 6, D. Gabbay and F. Guenthner (eds.), Kluwer Academic Publishers, Dordrecht, pp. 287-393, 2002.

[30] I. Urbas, Dual-intuitionistic logic, Notre Dame Journal of Formal Logic 37, pp. 440-451, 1996.

[31] H. Wansing, Constructive negation, implication, and co-implication, Journal of Applied Non-Classical Logics 18, pp. 341-364, 2008.

                                    Received 30 April 2019

Design of an MTJ-based nonvolatile multi-context ternary content-addressable memory

Naoya Onizawa

Research Institute of Electrical Communication, Tohoku University, Japan
naoya.onizawa.a7@tohoku.ac.jp

Ren Arakawa

Research Institute of Electrical Communication, Tohoku University, Japan
ren.arakawa.t3@dc.tohoku.ac.jp

Takahiro Hanyu

Research Institute of Electrical Communication, Tohoku University, Japan
hanyu@riec.tohoku.ac.jp

Abstract

In this paper, we present for the first time a design of multi-context ternary content-addressable memory (MC-TCAM) based on CMOS/magnetic tunnel junction (MTJ) devices. TCAMs are one of associative memories that realize fast search operations, where the applications are IP lookup engines and memory-based approximate computing. The multi-context capability is realized using a shared comparison circuit with multiple MTJ-based nonvolatile storage elements, resulting in a high memory density and a low static power dissipation. In addition, a CMOS-based cross-coupled circuit makes it possible to realize fast switching speed and low power dissipation, where the area overhead is negligible because of the shared structure. The proposed nonvolatile MC-TCAM with a 128 x 64 array of four contexts is designed using TSMC 65-nm CMOS and an MTJ model, which achieves a 2.6 x faster search speed and a 90−97% search-energy reduction in comparison with a single-context nonvolatile TCAM.

1   Introduction

Content-addressable memories (CAMs) are known as an associative memory [1] that handle input data instead of an input address used in traditional index memories, such as static random access memories (SRAMs) and dynamic random access memories (DRAMs). CAMs realize a parallel search operation of input data comparing with stored data, where binary CAMs store binary data and ternary CAMs store binary data with wildcard. The applications of CAMs include a cache [2], an intrusion detector [3], and an IP lookup engine [4]. In addition, CAMs have been recently applied for memory-based computing as lookup tables (LUTs) are realized using the stored data [5]. In [6], frequent operations in floating-point units are replaced by CAMs in Big Data applications [7, 8] in order to reduce the energy dissipation.

However, CAMs suffer from a high standby power issue and low area utilization. For example, a CMOS-based TCAM requires 16 transistor per cell as opposed to 6 transistors in SRAMs [9]. Recently, nonvolatile CAMs have been presented that significantly reduce the standby power dissipation with high density. Among nonvolatile devices, such as phase-change RAM (PCRAM) [10] and resistive RAM (RRAM) [11], magnetic tunnel junction (MTJ) [12] are often selected because of high-speed switching and high write endurance [13–15]. For further improving the area efficiency, a multi-context binary CAM using CMOS/MTJ devices was presented, where multiple storage elements are embedded in the same comparison circuit for search operations [16], but the function is limited to a binary comparison.

In this paper, a nonvolatile multi-context (MC) ternary CAM using CMOS/MTJ devices is introduced. The proposed MC-TCAM cell with N contexts is designed based on a single-context TCAM [15] with a structure of 10 transistors and 4 MTJs (10T-4MTJ). The proposed $(17/N + 4)T$-4MTJ cell uses more transistors ($17T$) for a more energy-efficient comparison circuit, reducing the search delay and the power dissipation. As the comparison circuit is shared for multiple storage elements, the cell area can be smaller than the single-context TCAM cell in case of large N, where the performance degradation due to large N is negligibly small. In addition, a dual-writing scheme is proposed that realizes a one-cycle writing to 4 MTJ devices instead of two cycles of the conventional single-context nonvolatile TCAM. As a design example, the proposed MC-TCAM with a 128 x 64 array of four contexts is designed using TSMC 65-nm CMOS and an MTJ model [17]. It achieves a 2.6 x faster search speed, a 90−97% lower search energy, and a 70% lower write energy in comparison with the single-context TCAM. The performance overhead due to larger number of contexts is also evaluated.

The rest of this paper is follows. Section 2 reviews the single-context TCAM. Section 3 describes the proposed MC-TCAM architecture and the $(17/N+4)T$-4MTJ cell with the dual-writing scheme. Section 4 evaluates the proposed MC-TCAM and compares with the single-context TCAM and related works. Section 5 concludes this paper.

2 Review of single-context nonvolatile TCAM

2.1 Overview of TCAM

Fig. 1 shows a TCAM architecture [1] containing 3 words with 4 bits. TCAMs realize fast and parallel search operations. Each word contains several TCAM cells that store '0', '1', or 'wildcard (X)'. A search data is compared with all the words in parallel to find which word matches. In this example, the search data of '0100' matches the first row of '01X0'. The output of the first row changes to "match" while that of the other rows remain "mismatch". Finally, the data of 'A' is read using this match signal. Basically, TCAMs have been designed for single-context search operations, where each TCAM cell contains the 1-bit information [5].

2.2 Nonvolatile CMOS/MTJ-based cell for single-context TCAMs

A TCAM cell can be designed using CMOS (volatile) devices and/or nonvolatile devices, such as PCRAM [10], ReRAM [11], and MTJ devices [12]. Fig. 2 shows a schematic of MTJ device with this symbol. The MTJ device is realized between an upper electrode (UE) and a lower electrode (LE) that can be stacked over a CMOS layer. It contains a free layer, a tunnel-barrier layer, and a fixed layer. The free layer is at one of two states depending on the spin direction that can be changed using a current, I_{MTJ}. If the spin direction of the free layer is the same of the fixed layer, the MTJ device is at a parallel state with low resistance (R_L). Otherwise, it is at an anti-parallel state with high resistance (R_H). As the resistance remains without a power supply, the MTJ devices can be used as nonvolatile storage elements. Due to this unique feature, the MTJ devices have been applied for several applications, such as nonvolatile flip-flops, processors, search engines, and true random number generators [18–23].

Fig. 3 shows a conventional nonvolatile $10T$-4MTJ cell for single-context TCAMs [15]. A search bit is represented by a dual-rail search line signal, (SL, $\overline{SL}$), where '0' and '1' are represented by (low, high) and (high, low), respectively. A stored bit contains '0', '1', or 'wildcard (X)' using four resistances: R_A, R_B, R_C, and R_D. There are two operations: search and write. In the search operation at the precharge phase, pre is low to precharge D and $\overline{D}$ to high. At the evaluate phase, the search

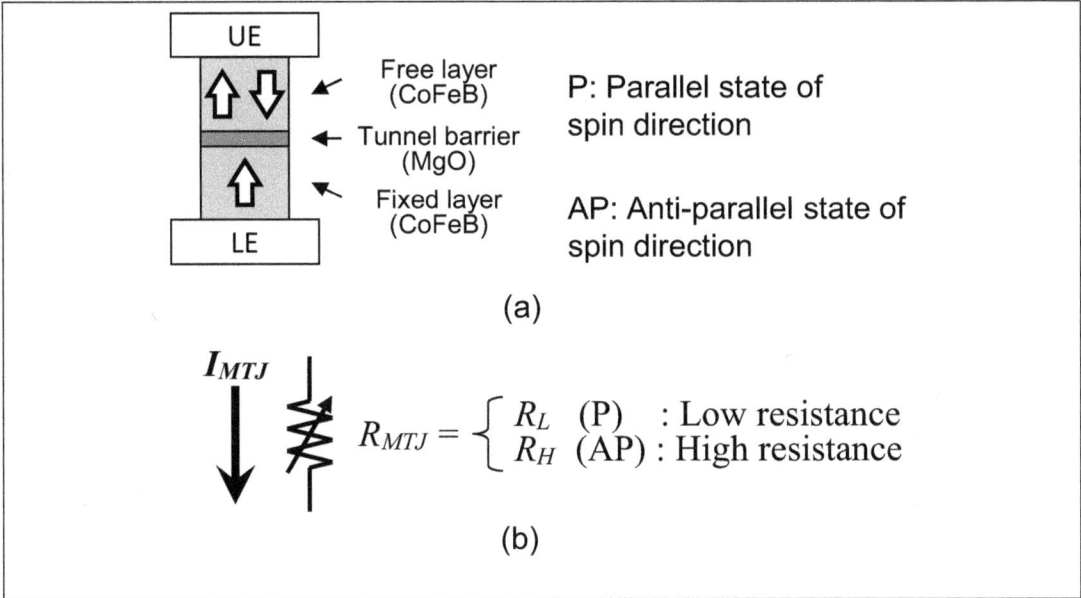

Figure 2: MTJ device: (a) schematic and (b) symbol.

bit is compared with the stored bit. If the search bit matches, match line (ML) remains high. Otherwise, ML changes to low.

In the write operation, four MTJ devices are written using a current for two cycles with $write$ = high. For example, when 'Data 0' is written, SL changes to high and $\overline{SL}$ remains low with $(BL, \overline{BL})$ = (high, low), generating a current signal through R_A and R_D at the first cycle. This results in the switching of R_A to R_H and that of R_D to R_L. At the second cycle, SL changes to low and $\overline{SL}$ changes to high with $(BL, \overline{BL})$ = (low, high), resulting in the switching of R_B to R_L and that of R_C to R_H.

3 Nonvolatile multi-context (MC) TCAM using CMOS/MTJ devices

3.1 Overall architecture

Fig. 4 shows the proposed MC-TCAM architecture with five contexts. As opposed to the conventional single-context TCAMs, each TCAM cell contains several context bits selected by a multiplexer. The search data is compared with one of selected contexts in parallel. By switching to another context, a different search operation

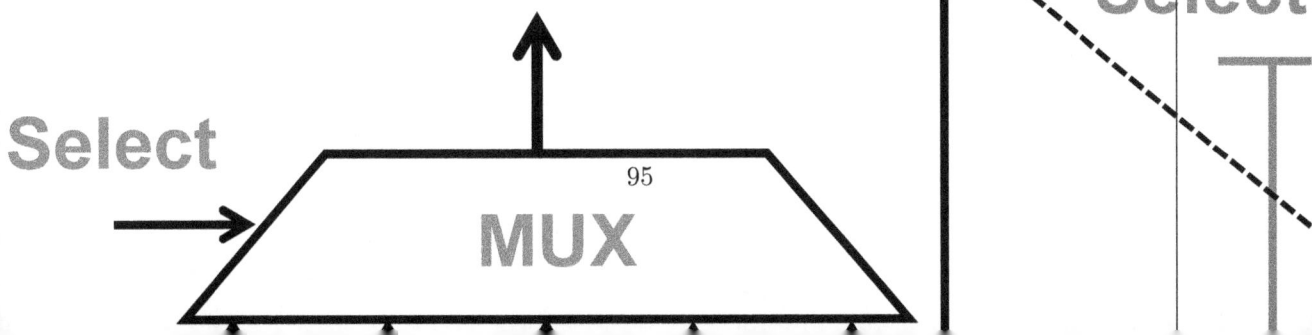

Select

95

MUX

Select

Select

can be immediately realized without writing another context to TCAM cells.

In [6], four different contexts are used to realize memory-based approximate computing. In this application, four single-context TCAMs are separately designed, resulting in a low area utilization. Currently, a multi-context binary CAM was presented in [16]; however, a multi-context TCAM has not been presented. This is the first time that multi-context TCAM is presented, which can improve the area utilization of the memory-based approximate computing.

3.2 MC-TCAM cell based on dual-writing scheme

Fig. 5 shows the proposed MC-TCAM cell with N contexts. The multi-context cell is designed based on the $10T$-4MTJ cell, where the same data representation is used in Fig. 3 (b). There are four MTJ blocks that represent R_A, R_B, R_C, and R_D. Each block selects one of N MTJ devices using selection signals ($S1$, $S2$, ... SN). When N is increased, four MTJ devices and four selection transistors per context are increased while the other circuit blocks remain the same. Therefore, the normalized cell size for 1-bit stored information is $(17/N + 4)T$-4MTJ.

Compared with the $10T$-4MTJ cell, transistors for SL and $\overline{SL}$ are separated from the writing paths to the MTJ devices. The separated structure reduces the transistor widths, reducing the power dissipation of search operations. In addition, the cross-coupled keeper is designed using both PMOS and NMOS transistors to increase a voltage difference between D and $\overline{D}$, which immediately cuts off a search current through the keeper for low-power search operations.

Fig. 6 shows a search operation of the MC-TCAM cell with a context selected by $S1$. There are precharge and evaluate phases as well as the $10T$-4MTJ cell. At the precharge phase shown in Fig. 6 (a), *pre* is low to precharge D and $\overline{D}$ to the same voltage level. At the evaluate phase shown in Fig. 6 (b), *pre* is high and a search line (SL and $\overline{SL}$) is active. In this example, a search bit of '1' is exhibited with $SL = High$ and $\overline{SL} = Low$. As a bit of '0' is stored with $R_A = R_H$ and $R_D = R_L$, the search operation is 'Mismatch', changing ML to low. By switching to another context using a different selection signals ($S1$, $S2$, ... SN), the search operation could be 'Match' with the same search bit.

Fig. 7 shows a write operation of the MC-TCAM cell. As opposed to the two-cycle write operation of the conventional $10T$-4MTJ cell, the proposed MC-TCAM cell writes the four MTJ devices for one cycle based on a dual-writing scheme. In the dual-writing scheme, two transistors for WE is active using four BLs, generating two current signals, I_{BC} and I_{AD}. Each current signal passes through three transistors and two MTJ devices as well as the $10T$-4MTJ cell. As the number of transistors and MTJ devices is increased in this write-current path, transistors for large widths

are required to maintain the same amount of current for writing MTJ devices. As a result, the proposed dual-rail writing scheme reduces the write cycle by half while maintaining the similar transistor sizes for writing in comparison with the single-context $10T$-4MTJ TCAM.

4   Evaluation

4.1   Simulated waveforms

The proposed MC-TCAM is designed using TSMC 65-nm CMOS and an MTJ model [17]. The TCAM array contains 64 words with 128 bits used in [9]. It is simulated using HSPICE with $VDD = 1.0V$ and 27°C. The resistances of the MTJ model are R_L of 763 Ω and R_H of 1,963 Ω.

Fig. 8 (a) shows simulated waveforms of search operations of the proposed MC-TCAM with the number of contexts $(N) = 4$. The MC-TCAM operates at 1.0 GHz with precharge (P) and evaluate (E) phases, which searches four times with two contexts in this example. In the context 1, the search data '0' matches the stored data, while the search data '1' mismatches. After switching to the context 2, the search data '1' matches because the stored data switches from '0' to '1'. Note that the switching delay of S1 and S2 depends on the memory size of the MC-TCAM. It is demonstrated that the context switching is immediately realized within a clock cycle.

Fig. 8 (b) shows simulated waveforms of write operations of the proposed MC-TCAM. Note that the write time of MTJ devices are longer than the read time [12]. Hence, in this paper, the write operation is at 100 MHz, while the search operation is at 1 GHz. First, '0' is stored using four MTJ devices as summarized in Fig. 5 (b). Then, these MTJ devices are written by I_{BC} and I_{AD} based on the dual-writing scheme. After the 10-ns writing, all the four MTJs are switched, resulting in storing '1'.

4.2   Comparisons with single-context CMOS/MTJ TCAM

The proposed MC-TCAM with $N = 4$ is compared with the single-context TCAM ($10T$-4MTJ) [15]. The $10T$-4MTJ TCAM is designed using the same technology and array size as the proposed MC-TCAM. Table 1 summarizes the comparisons of the search operation. The energy dissipation is categorized as ML (match line), SL (search line), cell, and other, such as context selection signals and bit lines. The energy dissipations of the proposed TCAM can be changed depending on a probability of context switching per search operation (P_{CS}).

101

Figure 8: Simulated waveforms of the proposed MC-TCAM: (a) search operations, where the match result is changed using the same search data because the context switches, and (b) write operations, where four MTJ devices are written based on

		Single-context [15]	This work (multi-context)	
			$P_{CS} = 0\%$	$P_{CS} = 100\%$
Search delay [ps]		326	126	124
Search energy [fJ/bit]	ML	0.036	0.12	0.12
	SL	2.10	0.240	0.237
	Cell	113	1.94	2.00
	Other	0.162	1.28	9.32
	Total	115	3.59	11.7

Table 1: Comparisons of search operation in 128 x 64 TCAMs.

The ML energies are small portions of the total. In terms of the SL energies, the single-context TCAM consumes around 9 times larger because of the cell structure. The single-context TCAM cell uses 10 transistors that shares transistors for search and write, while the proposed cell uses 17 transistors for separated search and write transistors. The search-write shared structure needs to increase the widths of the transistors because MTJ devices require large currents for writing, such as 100 μA. In contrast, the proposed separated structure uses smaller widths, reducing the search energy.

In terms of the cell energies, the single-context TCAM consumes around 57 times larger than the proposed cell. The cross-coupled keeper is designed using only PMOS transistors in the single-context cell, while the proposed cell uses both PMOS and NMOS transistors as shown in Fig. 9. This reduced-complexity design of the single-context cell causes a steady current during search operations, resulting in a large energy dissipation of cell. In contrast, the proposed CMOS-based cross-coupled keeper immediately cuts off the switching current because of a full voltage swing of the outputs, which reduces the energy dissipation. In addition, it realizes fast switching of transistors for ML, reducing the search delay by 62%. In the proposed cell, the energy dissipation (other) is larger when P_{CS} is larger because the energy dissipation of the context selection signals are frequently changed. As a result, the proposed MC-TCAM reduces the total energy dissipation of searching by 90–97% in comparison with the single-context TCAM depending on P_{CS}.

Table 2 summarizes the performance comparisons of the write operation. The single-context cell requires two cycles for writing, while the proposed cell reduces the cycle to 1 because of the dual-writing scheme, reducing the write time by half. As the search-write shared structure is used in the single-context cell, transistors of large widths for SL are required with a boosted supply voltage of 1.5 V, resulting in the large energy dissipation of SL. In contrast, the proposed search-write separated

I_{VDD} VDD

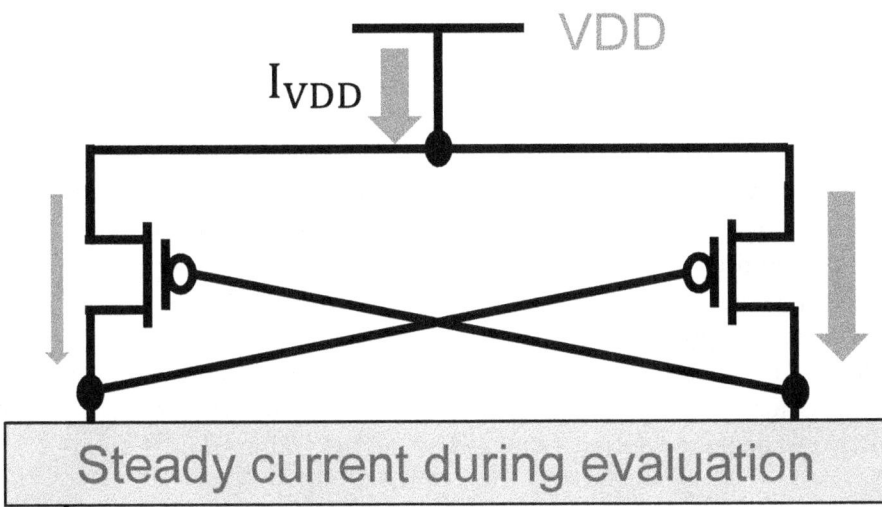

Steady current during evaluation

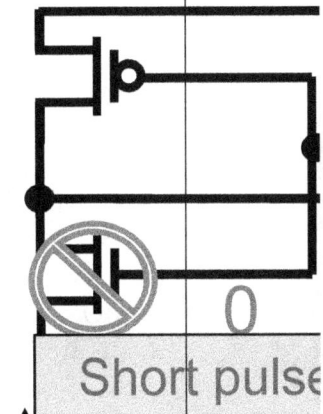

0

Short pulse

103

200

VDD

P E P

		Single-context [15]	This work (multi-context)
Supply voltage [V]		1.0/1.5	1.0
Number of write cycles/bit		2	1
Write time [ns]		20	10
Write energy [pJ/bit]	ML	0.0007	0.0005
	SL	19.0	0.0262
	Cell	1.5	0.088
	Other	19.2	11.8
	Total	39.8	11.9

Table 2: Comparisons of write operation in 128 x 64 TCAMs.

structure enables us to use a normal voltage of 1.0 V for writing.

In total, the proposed MC-TCAM reduces the total energy dissipation of writing by 70%. As a result, the proposed TCAM achieves a 2.6 x faster search speed, a 90−97% lower search energy, and a 70% lower write energy than the single-context TCAM. As described in this subsection, the proposed cell exploits larger number of transistors, resulting in lower search delay time and lower energy dissipations than the single-context cell. However, this area overhead can be reduced when the number of contexts (N) is increased. As shown in Fig. 10 (a), the normalized area/bit is smaller when N is larger than 3. In addition, the performance overhead due to larger N is small. For example, the search delay and energy at $N = 16$ are respectively 4.3% and 5.3% larger than that at $N = 2$ as shown in Fig. 10 (b).

4.3 Related works and discussion

The proposed MC-TCAM is compared and discussed with related works that include a SRAM-based TCAM [9], a CMOS/MTJ-based TCAM [15], and a CMOS/MTJ-based multi-context binary CAM [16] as summarized in Table 3. The SRAM-based TCAM realizes the lowest search energy dissipation because a precharge/evaluate operation per cell is not required every clock cycle, unlike MTJ-based TCAMs. However, the TCAM cell requires 16 transistors, resulting in a large cell area and a large static power dissipation as the storage elements are volatile. The single-context nonvolatile TCAM can significantly reduce the static power dissipation by the nonvolatility of the MTJ devices. Note that MTJ devices are implemented over a CMOS layer that can ignore the area of MTJ devices in terms of cell area [24]. In [15], it is reported that the single-context nonvolatile TCAM cell is smaller than a SRAM-based TCAM cell.

	[9]	[15]	[16]	This work
Process	65-nm CMOS	65-nm CMOS +MTJ	28-nm CMOS +MTJ	65-nm CMOS +MTJ
Feature	**Ternary**	**Ternary**	Binary	**Ternary**
Context	Single	Single	**Multiple**	**Multiple**
Search delay [ps]	650	326	110	126
Search energy [fJ/bit]	0.54	120	3.2	3.6
Write energy [pJ/bit]	N/A	40	N/A	12
Static power	Yes	**No**	**No**	**No**
Normalized cell area/bit	$16T$	$10T$ +4MTJ	$(19/N+2)T$ +2MTJ	$(17/N+4)T$ +4MTJ

Table 3: Performance comparisons with related works, where T is the number of transistors and N is the number of contexts.

The multi-context nonvolatile binary CAM realizes multi-context search operations while the function is limited to binary comparisons. Note that several applications require ternary operations, such as IP lookup [4] and memory-based approximate computing [6]. In comparison with the multi-context binary CAM, the proposed MC-TCAM using an older CMOS technology achieves similar search delay time and energy dissipation while realizing the ternary search operation. The reason of the comparable performance derives from the circuit structure of the TCAM cells, where the number of transistors from VDD to GND determines the worst-case delay. The cell in [16] contains six transistors from VDD to GND while the proposed TCAM cell contains only four transistors, which can enable us to design more efficient cell structure.

5 Conclusions

In this paper, we have presented the nonvolatile multi-context TCAM using CMOS/ MTJ devices. The proposed $(17/N+4)T$-4MTJ cell realizes the multi-context search capability while increasing the search speed and search energy in comparison with the single-context TCAM cell ($10T$-4MTJ). The performance improvement derives from the energy-efficient comparison circuit using larger number of transistors, while the area overhead can be negligible in case of large number of contexts (N). In addition, the proposed dual-writing scheme realizes the one-cycle writing for 4 MTJ devices while two cycles are required in the conventional TCAM, resulting in lower write time and write energy. The proposed MC-TCAM with a 128 x 64 array of four contexts is designed using TSMC 65-nm CMOS and an MTJ model that achieves a 2.6 x faster search speed, a 90−97% lower search energy, and a 70% lower write energy than the single-context TCAM.

In the future prospect, the proposed MC-TCAM could be applied for memory-based approximate computing [6] that uses several TCAMs with different contexts. Using our technique, separated TCAMs with different contexts could be replaced by the MC-TCAM, reducing the area overhead.

Acknowledgment

This work was supported by JSPS KAKENHI Grant Numbers JP16H06300, and VLSI Design and Education Center (VDEC), The University of Tokyo with the collaboration with Synopsys Corporation.

References

[1] K. Pagiamtzis and A. Sheikholeslami. Content-addressable memory (CAM) circuits and architectures: a tutorial and survey. *IEEE Journal of Solid-State Circuits*, 41(3):712–727, 2006.

[2] Perng-Fei Lin and J.B. Kuo. A 1-V 128-kb four-way set-associative cmos cache memory using wordline-oriented tag-compare (WLOTC) structure with the content-addressable-memory (CAM) 10-transistor tag cell. *IEEE Journal of Solid-State Circuits*, 36(4):666 –675, Apr. 2001.

[3] Chao-Ching Wang, Chieh-Jen Cheng, Tien-Fu Chen, and Jinn-Shyan Wang. An adaptively dividable dual-port BiTCAM for virus-detection processors in mobile devices. *IEEE Journal of Solid-State Circuits*, 44(5):1571 –1581, May 2009.

[4] Nen-Fu Huang, Whai-En Chen, Jiau-Yu Luo, and Jun-Min Chen. Design of multi-field IPv6 packet classifiers using ternary CAMs. In *Proc. Global Telecommunications Conference, 2001*, volume 3, pp. 1877 - 1881, 2001.

[5] R. Karam, R. Puri, S. Ghosh, and S. Bhunia. Emerging trends in design and applications of memory-based computing and content-addressable memories. *Proceedings of the IEEE*, 103(8):1311–1330, Aug 2015.

[6] M. Imani, S. Patil, and T. Šimunić Rosing. Approximate computing using multiple-access single-charge associative memory. *IEEE Transactions on Emerging Topics in Computing*, 6(3):305–316, July 2018.

[7] A. Katal, M. Wazid, and R. Goudar. Big data: Issues challenges tools and good practices. In *6th Int. Conf. Contemporary Comput.*, 2013.

[8] C. Ji, Y. Li, W. Qiu, U. Awada, and K. Li. Big data processing in cloud computing environments. In *12th Int. Symp. Pervasive Syst. Algorithms Networks*, pages 17–23, 2012.

[9] W. Choi, K. Lee, and J. Park. Low cost ternary content addressable memory using adaptive matchline discharging scheme. In *2018 IEEE International Symposium on Circuits and Systems (ISCAS)*, pages 1–4, May 2018.

[10] M. Wuttig and N. Yamada. Phase-change materials for rewriteable data storage. *Nature Materials*, 6(11):824 – 832, 2007.

[11] M. F. Chang, C. W. Wu, C. C. Kuo, S. J. Shen, K. F. Lin, S. M. Yang, Y. C. King, C. J. Lin, and Y. D. Chih. A 0.5V 4Mb logic-process compatible embedded resistive RAM (ReRAM) in 65nm CMOS using low-voltage current-mode sensing scheme with 45ns random read time. In *2012 IEEE International Solid-State Circuits Conference*, pages 434–436, Feb 2012.

[12] S. Ikeda et al. A perpendicular-anisotropy CoFeBMgO magnetic tunnel junction. *Nature Materials*, 9:721 – 724, 2010.

[13] S. Matsunaga, K. Hiyama, A. Matsumoto, S. Ikeda, H. Hasegawa, K. Miura, J. Hayakawa, T. Endoh, H. Ohno, and T. Hanyu. Standby-power-free compact ternary content-addressable memory cell chip using magnetic tunnel junction devices. *Applied Physics Express*, 2:023004, feb 2009.

[14] S. Matsunaga, S. Miura, H. Honjou, K. Kinoshita, S. Ikeda, T. Endoh, H. Ohno, and T. Hanyu. A 3.14 um2 4T-2MTJ-cell fully parallel TCAM based on nonvolatile logic-in-memory architecture. In *2012 Symposium on VLSI Circuits (VLSIC)*, pages 44–45, June 2012.

[15] B. Song, T. Na, J. P. Kim, S. H. Kang, and S. Jung. A 10T-4MTJ nonvolatile ternary CAM cell for reliable search operation and a compact area. *IEEE Transactions on Circuits and Systems II: Express Briefs*, 64(6):700–704, June 2017.

[16] E. Deng, L. Anghel, G. Prenat, and W. Zhao. Multi-context non-volatile content addressable memory using magnetic tunnel junctions. In *2016 IEEE/ACM International Symposium on Nanoscale Architectures (NANOARCH)*, pages 103–108, July 2016.

[17] N. Sakimura, R. Nebashi, Y. Tsuji, H. Honjo, T. Sugibayashi, H. Koike, T. Ohsawa, S. Fukami, T. Hanyu, H. Ohno, and T. Endoh. High-speed simulator including accurate MTJ models for spintronics integrated circuit design. In *2012 IEEE International Symposium on Circuits and Systems*, pages 1971–1974, May 2012.

[18] N. Sakimura, T. Sugibayashi, R. Nebashi, and N. Kasai. Nonvolatile magnetic flip-flop for standby-power-free SoCs. *IEEE Journal of Solid-State Circuits*, 44(8):2244–2250, Aug 2009.

[19] N. Sakimura, Y. Tsuji, R. Nebashi, H. Honjo, A. Morioka, K. Ishihara, K. Kinoshita, S. Fukami, S. Miura, N. Kasai, T. Endoh, H. Ohno, T. Hanyu, and T. Sugibayashi. A 90nm 20MHz fully nonvolatile microcontroller for standby-power-critical applications. In *2014 IEEE International Solid-State Circuits Conference Digest of Technical Papers (ISSCC)*, pages 184–185, Feb 2014.

[20] T. Kawahara, R. Takemura, K. Miura, J. Hayakawa, S. Ikeda, Y. M. Lee, R. Sasaki,

108

Y. Goto, K. Ito, T. Meguro, F. Matsukura, H. Takahashi, H. Matsuoka, and H. Ohno. 2 Mb SPRAM (spin-transfer torque ram) with bit-by-bit bi-directional current write and parallelizing-direction current read. *IEEE Journal of Solid-State Circuits*, 43(1):109–120, Jan 2008.

[21] H. Jarollahi, N. Onizawa, V. Gripon, N. Sakimura, T. Sugibayashi, T. Endoh, H. Ohno, T. Hanyu, and W. J. Gross. A nonvolatile associative memory-based context-driven search engine using 90 nm CMOS/MTJ-hybrid logic-in-memory architecture. *IEEE Journal on Emerging and Selected Topics in Circuits and Systems*, 4(4):460–474, Dec 2014.

[22] J. Diguet, N. Onizawa, M. Rizk, J. Sepulveda, A. Baghdadi, and T. Hanyu. Networked power-gated MRAMs for memory-based computing. *IEEE Transactions on Very Large Scale Integration (VLSI) Systems*, 26(12):2696–2708, Dec 2018.

[23] S. Mukaida, N. Onizawa, and T. Hanyu. Design of a low-power MTJ-based true random number generator using a multi-voltage/current converter. In *2018 IEEE 48th International Symposium on Multiple-Valued Logic (ISMVL)*, pages 156–161, May 2018.

[24] T. Hanyu, T. Endoh, D. Suzuki, H. Koike, Y. Ma, N. Onizawa, M. Natsui, S. Ikeda, and H. Ohno. Standby-power-free integrated circuits using MTJ-based VLSI computing. *Proceedings of the IEEE*, 104(10):1844–1863, Oct 2016.

 Received 30 April 2019

www.ingramcontent.com/pod-product-compliance
Lightning Source LLC
Chambersburg PA
CBHW081332090426

42737CB00017B/3106